Semiarid Soil and Water Conservation

Author
Herman J. Finkel, Ph.D.
Director
Finkel & Finkel, Consulting Engineers
Yoqneam, Israel

Contributors
Moshe Finkel
Agricultural Engineer
Director
Finkel & Finkel, Consulting Engineers
Yoqneam, Israel

Ze'ev Naveh, Ph.D.
Professor of Ecology
Faculty of Agricultural Engineering
Technion, Israel Insitute of Technology
Haifa, Israel

CRC Press, Inc.
Boca Raton, Florida

Library of Congress Cataloging-in-Publication Data

Finkel, Herman J.
  Semiarid soil and water conservation.
  Includes bibliographies and index.
  1. Soil conservation.   2. Water conservation.
3. Arid regions agriculture. I. Title.
S623.F48 1986   631.4'5   85-29157
ISBN 0-8493-6112-5

This book represents information obtained from authentic and highly regarded sources. Reprinted material is quoted with permission, and sources are indicated. A wide variety of references are listed. Every reasonable effort has been made to give reliable data and information, but the author and the publisher cannot assume responsibility for the validity of all materials or for the consequences of their use.

Direct all inquiries to CRC Press, Inc., 2000 Corporate Blvd., N.W., Boca Raton, Florida, 33431.

International Standard Book Number 0-8493-6112-5

Library of Congress Card Number 85-29157
Printed in the United States

THE AUTHORS

Herman J. Finkel, Ph.D., is Director of Finkel & Finkel, Consulting Engineers, and Emeritus Professor of Agricultural Engineering of the Technion-Israel Institute of Technology, Haifa, Israel.

Dr. Finkel is a graduate of the Department of Agricultural Engineering of the University of Illinois (1940) and holds a Ph.D. from the Faculty of Agriculture of the Hebrew University, Jerusalem. After years of experience in the U.S. with the U.S. Corps of Engineers, the U.S. Soil Conservation Service, and private consulting engineering firms, he moved to Israel where he became the chief engineer of the newly formed Israel Soil Conservation Service. This work emphasized water resource development and irrigation. In 1952, he founded the Faculty of Agricultural Engineering at the Technion-Israel Institute of Technology, and served as its head and dean intermittently for 24 years. During this time he taught and did research in the fields of irrigation and soil and water. From 1969 through 1972 he served as Academic Vice-President for the Technion. During this period, Dr. Finkel maintained an active consulting practice and served as an expert in irrigation for FAO, other international agencies, and private consulting firms in 24 countries of Latin America, the Caribbean, Africa, the Middle East, and the Far East. This activity was divided among design, project evaluation, and training of local staff. He also served on a number of missions of UNESCO and the World Bank on educational planning for the developing countries, as an expert on agricultural and technical education. Since 1972, he has been the head of the consulting firm of Finkel & Finkel. This firm has done design work in Israel as well as in Iran, the Caribbean, and Latin America.

Dr. Finkel is the author of numerous articles both in his profession and in the more distant fields of history and archaeology. He recently edited the *CRC Handbook of Irrigation Technology.*

Ze'ev Naveh is Professor of Landscape Technology at the Technion-Israel Institute of Technology in both the Faculty of Agricultural Engineering and Architecture and Town Planning.

He has had long-term experience in management and improvement of Mediterranean uplands for pasture and multipurpose afforestation, also in the drier parts of East Africa. He serves as an international consultant in several semiarid zones.

Moshe Finkel, Agricultural Engineer, is a Director of Finkel and Finkel, Consulting Engineers, Haifa, Israel, and serves as a consultant for FAO and other international agencies on water harvesting in semiarid zones of Africa.

TABLE OF CONTENTS

Chapter 1

INTRODUCTION

Herman J. Finkel

I. HISTORICAL

The hazards of accelerated soil erosion were first studied by a few pioneer conservationists, notably Lowdermilk and Bennett. Lowdermilk, in his work on famine relief in China after World War I, was deeply impressed by the millions of tons of fertile soil which were annually washed down the Yangtze River (known as China's Sorrow), leaving the upland watersheds bare, infertile, and ravaged by gullies. Upon his return to the U.S. he established, at San Dimas, Calif., the first experimental watershed studies for the quantitative evaluation of rates of soil erosion. He and a few other scientists, mainly foresters, sounded warnings that the fertile topsoil of the U.S. stood in great danger of being washed away by the rains at an increasingly accelerated rate.

In the Great Depression of the 1930s, President Roosevelt inaugurated an emergency program of public works to relieve unemployment. He created governmental agencies to prepare plans for the constructive use of surplus labor. One of these was the Soil Conservation Service (SCS), which undertook soil erosion and land use surveys for the entire country. This was followed by a vigorous program of planning and extension to establish conservation measures in the field. Another agency was the Civilian Conservation Corps (CCC), which took the unemployed youth off the city streets and gave them healthy, creative work in the field camps. The main activities of the CCC were reforestation and the construction of structures for gully control, water storage, flood water diversion, and wind erosion control. The methods and practices of these two agencies provided a foundation for soil conservation programs in the U.S., and eventually, all over the world.

The SCS, in its formative years, tried to become completely self-sufficient and independent of other governmental agencies. It conducted its own soil surveys, research, and extension work. These were functions which had long been in the province of other, established agencies. The justification for this apparant duplication of effort was that the attack on the soil erosion problem required an integrated approach in which all the activities must be closely interrelated and handled by a large and multidisciplined field staff of conservationists and farm planners, with technical support from a smaller number of specialists.

This approach was quite different from the conventional division of authority among numerous governmental bureaus. It speeded up operations tremendously, and resulted in impressive achievements in the field. Not the least of these was the rapid provision of working plans for the army of young people in the conservation camps waiting to do some useful work. On the other hand it created jealousy and rivalry with the older, established branches of the U.S. Department of Agriculture (USDA), which had some negative results. In those early "heady" days, the energy and enthusiasm of the dedicated SCS field staff was the "priceless ingredient". It compensated, to a certain extent, for the institutional problems of the so-called "empirialistic approach", i.e., ignoring or supplanting the traditional, old-line agencies of the USDA. This has all changed now, and it will be a matter for historians to judge whether true progress in vital public projects is better achieved by conventional or unconventional modes of organization.

During the first few decades of its existence, the SCS received and trained a great many visiting technicians and administrators from all over the world. This influenced the rapid development of similar agencies in the other countries. The organizational structure, field procedures, and technical solutions developed in the U.S. served as a paradigm which was widely copied. However, indiscriminate technology transfer has certain hazards because of the unique conditions of the recipient countries, in climate, soils, agriculture, economics, social and institutional factors, and even cultural and humanistic aspects. Consequently, with the passage of time, many of the Third World countries are developing their own approaches and solutions which are more compatible with their needs and circumstances.

II. CHARACTERISTICS OF SEMIARID REGIONS

Semiarid regions have been defined in various ways. One of the earliest was developed by Thornthwaite in 1948, and is based upon the relationship between the monthly rainfall and the potential evapotranspiration. Another definition, proposed by de Martonne, is the aridity index,

$$\text{A.I.} = P/(t + 10) \tag{1}$$

where P is the average annual precipitation in millimeters and t the average annual temperature in degrees centigrade. Semiarid regions are defined as those where the aridity index falls between 10 and 20. The annual precipitation is modified by the temperature in such a way that a given precipitation represents a higher degree of aridity when accompanied by higher temperature, and vice versa. The potential evapotranspiration, or the maximum consumptive use of moisture by the plants, is greater at higher temperatures and consequently the annual moisture deficit is greater. Several other methods for estimating evapotranspiration from climatological data have been developed. These are discussed in detail in "Water requirements of crops and irrigation rates", in Finkel's *Handbook of Irrigation Technology* (CRC Press, 1982).

Maps of the world have been prepared by UNESCO (United Nations Educational, Scientific, and Cultural Organization), delineating the arid and semiarid zones. The moisture-deficient areas include the western part of the U.S., the eastern and southern shores of the Mediterranean, Iran, the Sahel south of the Sahara, a large part of the Indian subcontinent, much of the coastal strip of Australia, the western coast of South America, the summer rainfall region of southern Africa, and many other regions. These very extensive areas are deserving of special attention with respect to the problems of soil and water conservation.

From the point of view of planning for soil and water conservation, the definitions based upon long-term annual rainfall and temperature records are not sufficient. It is necessary to supplement them by a somewhat different approach. In the eastern part of the U.S. and in much of Europe, precipitation is fairly well distributed throughout the year, either as rain or snow. In many tropical countries rainfall is also quite abundant during most months of the year. However, in some regions, the rainfall may be interrupted by a definite dry season for a period of time. The Mediterranean climate, for example, has rain in the winter and a dry summer. Another example is the so-called monomodal tropical rainfall pattern found in West Africa.

There may also be two distinct dry spells, such as the bimodal equatorial distribution in India, where the monsoons divide the year into four distinct seasons: the cold season of the northeast monsoon from December to February, with some rainfall, the hot-dry season from March to May, the rains of the southwest monsoon from June to September, and the dry autumn from October to November.

The critical factor in every case is not the total annual precipitation, which on the long-term average, may show values equal to or higher than the estimated potential evapotranspiration. It is, rather, the length of the dry season which is important. This critical duration cannot be defined simply for all regions. It depends, partially, upon the antecedent precip-

Table 1
AVERAGE PRECIPITATION AND EVAPOTRANSPIRATION IN mm, KENYA EQUATOR STATION

Month	J	F	M	A	M	J	J	A	S	O	N	D	Total
P	36	36	70	165	140	125	165	200	110	55	60	60	1,222
ET	140	140	155	115	115	88	75	85	100	125	120	130	1,388

itation, the storage capacity of the soil in the depth of the root zone, and the water requirements of the major crops (assuming no irrigation). However, anything over 1 month of no precipitation during the normal growing season would justify classifying the region as semiarid. This is based upon long-term monthly averages; in any region there could be an infrequent occurrence of 1 or 2 months' drought without changing the classification of the climatic region. In the higher latitudes, and/or at higher altitudes, a regular dry season might coincide with temperatures too low for growing crops. Such a dry period would be less important.

An example of the inadequacy of the definition of aridity based upon long-term average annual precipitation may be seen in the Equator Station in Kenya (elev. = 2762 m). Records over the period from 1938 to 1962 indicate that the average annual precipitation is 1222 mm, and the average annual evapotranspiration is 1388 mm. The average monthly precipitation and evapotranspiration in millimeters are shown in Table 1.

It is seen that from October to March the evapotranspiration was higher than the precipitation by a considerable amount. From April through September the precipitation was much higher than the evapotranspiration. Yet, in terms of the annual totals, P and ET are more or less equal. The dry season of 6 months defines this location as semiarid, despite the fact that well over 1 m of rain falls in the average year.

The above criterion is based upon the long-term averages of monthly precipitation. There is, however, another form of drought which can result in a region being classified as semiarid. It is the highly erratic fluctuation from their long-term averages of specific monthly precipitation values. Thus, it is possible for the long-term monthly averages to indicate sufficient precipitation, while in certain years there may be an appreciable dry period which may occur cyclically or stochastically. This condition of unreliability of the rains, while well known to local farmers, may be difficult to reveal from an inspection of only the long-term monthly averages. The degree of variability is determined by calculating the probability of occurrence of various levels of precipitation in each month — this will be discussed in detail in Chapter 2. When such variability is taken into account, the delineation of arid and semiarid lands will become considerably extended.

III. UNIQUE PROBLEMS

A major problem of controlling water erosion in semiarid regions arises from the lack of coincidence between the rainfall which causes the erosion, and the vegetative cover which protects the soil surface. This is particularly true of cultivated cropland and heavily grazed pastures, but somewhat less so for tree crops. For the winter rainfall climate in subtropical regions, such as the eastern Mediterranean, the first rains usually come in October and fall upon the bare soil from which the summer crops have long been removed. Sometimes the soil is plowed right after the first rain, when the hard, dry clods become softer, and the winter grain is planted. In this condition, the soil is most exposed and highly vulnerable to the ravages of water erosion. Only after 45 to 60 days is a reasonable plant cover developed. Likewise, the thin, overgrazed summer pastures provide little protection in October and November, as only in December may sufficient growth develop to protect the soil. There are obviously two or three months, from October through mid-December, when soil erosion will be critical. The application of conservation measures must be planned accordingly.

In the winter rainfall regions of higher latitudes where the winter is too cold for plant growth, the grain is often planted in the fall, and may achieve a certain amount of growth with the first rains before the frost sets in. During the winter, the ice and snow effectively prevent soil erosion. After the spring thaw, the grain continues to grow and is harvested in early summer. The critical periods for soil erosion under this regime are a short time in the fall after plowing and before the frost, and during the last rains, which may coincide with the spring thaw, giving large amounts of runoff.

In the temperate and subtropical regions of summer rainfall there is generally a good coincidence between the rain and the vegetative cover, except at the very beginning of the rainy season before the cover is well established, and at the end of the rainy season when the cover may have been removed by harvesting or grazing before the last rains have ended. Where the rainfall is erratic and very unreliable the potentially critical periods for erosion are less defined and may occur at any time during the rainy season.

The basic approach to water erosion control in semiarid regions is, therefore, to superimpose a plot of the monthly rainfall upon the growth curves for the principal agricultural crops, in order to identify the period or periods when erosion may be most active. Suitable measures applied at the critical period will give greater benefit for less cost.

In this volume, the erosion and conservation measures discussed are, for the most part, those under unirrigated agriculture. The use of irrigation could cause significant changes in the growing seasons, and in the agricultural calendar, especially in the warmer climates where temperature is not a limiting factor. It is further noted that much of the material in this volume has been prepared with the developing countries of the so-called Third World in mind. In many of these countries there is a dearth of basic data, such as long-term hydrological records, detailed soil and topographic surveys, and experimental results for various types of erosion control measures. Some design procedures cannot be imitated or copied directly from those of the technologically more advanced countries. Consequently, emphasis will be placed, wherever possible, upon simple empirical methods of design, and approximate solutions within the limitations of the available data, technical possibilities, and financial resources of the Third World countries. Much of the numerical data and calculations will be presented in the metric system.

REFERENCES

1. **Hudson, N.**, *Soil Conservation*, Batsford Academic & Educational, London, 1981.
2. **Meigs, P.**, World distribution of arid and semi-arid homoclimates, in Reviews of Research on Arid Zone Hydrology, United Nations Educational, Scientific, and Cultural Organization, Paris, 1953.

Chapter 2

HYDROLOGY

Herman J. Finkel and Moshe Finkel

I. INTRODUCTION

Hydrologic studies are important in soil and water conservation planning to determine:

1. The rainfall characteristics for evaluating the erosion hazard
2. The relationships between rainfall and runoff
3. The peak discharges of surface flow for the design of hydraulic structures
4. The relationships between watershed characteristics and peak discharges so that measures can be taken to reduce these peaks

These objectives have been the chief preoccupation of hydrologists, primarily in the more humid regions. In the semiarid regions where the importance of water retention supercedes that of water disposal, the following additional objectives of determination may be listed:

5. The probabilities of subaverage precipitation or drought for purposes of estimating success in cropping
6. The probabilities of subaverage or minimum runoff for estimating the safe or reliable yield of a water harvesting scheme
7. The relationships between the watershed characteristics and total runoff, in order to increase water yield

It is unfortunate that in many semiarid regions of the Third World there are not sufficient long-term records of precipitation depth, duration, and intensity, and even fewer measurements of continuous discharge from the streams. For this reason, many of the sophisticated and advanced hydrologic methods are difficult to apply, since the mathematical models are much better than the data. Simplifications and empirical approximations may have to be used until the data collection networks are greatly improved. To a certain extent this approach will be expressed in this volume.

II. PRECIPITATION

Precipitation records are usually among the most available of climatological data in many countries. Analysis of these data is illustrated in Table 1. The table gives the monthly values of rainfall in millimeters for Kerman, a town in southern Iran. The record covers a period of 20 years and is listed according to the hydrologic year from October to September, rather than the calendar year. A number of interesting calculations and observations may be made from these data.

A. Annual Rainfall

The mean annual rainfall is simply the average of the annual totals, or 158 mm. The minimum annual total was 41.4 mm, and the maximum was 300.6 mm or over seven times as much. This is a very wide spread which calls for further analysis. The most simple and useful approach is to calculate the percentage probability of occurrence of different levels of annual rainfall. This may be done as shown in Table 2.

Table 1
MONTHLY PRECIPITATION (mm) IN KERMAN, IRAN

Year	Oct	Nov	Dec	Jan	Feb	Mar	Apr	May	Jun	Jul	Aug	Sep	Total
1968/69	—	3.3	29.8	30.6	34.8	8.8	6.2	5.0	—	—	—	—	118.5
1967/68	—	1.8	9.2	—	—	18.8	43.6	23.0	6.5	—	—	—	102.9
1966/67	—	—	4.5	12.5	60.3	2.0	37.6	3.0	—	—	—	—	120.9
1965/66	—	1.0	—	5.5	18.7	6.1	10.1	—	—	—	—	—	41.1
1964/65	—	—	7.2	100.2	13.7	8.0	2.0	—	—	2.8	—	—	133.9
1963/64	1.9	29.7	8.0	25.0	78.8	23.0	5.6	1.6	—	—	—	—	173.6
1962/63	—	0.3	17.0	0.6	68.7	42.8	6.6	20.6	—	—	—	0.1	156.7
1961/62	—	3.9	18.0	5.2	22.0	13.2	96.8	0.8	—	—	—	—	159.9
1960/61	—	—	—	14.1	12.1	30.8	25.5	17.6	—	—	—	—	100.1
1959/60	—	16.2	29.4	20.5	26.4	91.4	45.2	16.3	—	19.2	—	36.0	300.6
1958/59	1.2	1.5	8.9	25.7	3.0	4.8	6.5	43.5	—	0.4	—	—	95.5
1957/58	1.0	34.0	33.3	27.1	6.1	16.5	2.1	6.3	—	—	—	—	126.4
1956/57	—	—	26.7	103.4	12.1	31.1	61.6	26.3	7.8	—	0.5	—	269.7
1955/56	23.0	—	33.7	62.6	3.5	104.7	8.4	—	—	14.3	—	—	250.2
1954/55	4.0	27.8	37.5	38.6	21.0	82.4	5.1	18.2	11.1	1.5	—	—	247.2
1953/54	—	0.7	84.1	25.8	35.8	25.0	15.5	11.8	2.5	30.6	—	—	231.8
1952/53	—	3.5	25.5	4.4	37.2	28.6	59.6	48.2	—	—	—	—	207.8
1951/52	1.0	—	35.0	30.1	15.5	16.9	6.5	9.2	2.0	0.6	—	—	116.8
1950/51	—	17.5	7.5	19.8	60.1	61.4	12.4	22.3	0.3	—	—	1.9	203.2
1949/50				—	18.0	25.8	8.5	16.0	—	7.5	3.3	5.0	84.1

Table 2
PROBABILITY OF
OCCURRENCE OF ANNUAL
RAINFALL IN KERMAN

Equal to or greater than (mm)	No. of years	Probability (% = or >)
50	19	95
100	17	85
150	11	55
158	10	50
200	7	35
250	3	15
300	1	5

In simple terms this means that a rainfall of at least 50 mm can be expected every year. On an average of once in 2 years the rainfall may equal or exceed 158 mm, once in 5 years it may reach 250 mm, and once in 30 years it may reach or exceed 300 mm. Of course, it is understood that these recurrence intervals are expected averages over a long time and will not necessarily occur at periodic intervals. Because of the popular misunderstanding of the expression "recurrence interval", it is better to use the percentage probability that an event of a given magnitude may occur in any year.

The above analysis can be of value in predicting the effectiveness of water conservation measures such as a surface reservoir or a system of water-harvesting bunds. If the runoff is assumed to be roughly proportional to the precipitation (an assumption which may have to be modified in specific cases), it can be concluded that the minimum safe yield in the Kerman area should be based upon a rainfall of 50 mm/year, and about three times as much could be expected on the average of once in 2 years. However, the structure may be flooded with five times as much runoff on the average of once in 5 years and six times as much once in

Table 3
PROBABILITY OF OCCURRENCE OF DROUGHTS OF VARIOUS DURATIONS AT KERMAN

Duration in mon. (= or >)	No. of occurrences	Probability (%)	Frequency
3	20	100	1
4	16	80	1.25
5	12	60	1.67
6	8	40	2.5
7	5	25	4
8	2	10	10
9	1	5	20

20 years. The degree of protection to be provided for the less frequent large flows is a matter of economics and will be discussed in a subsequent chapter.

B. Drought Period

The durations and frequencies of the droughts are matters of interest. In the Kerman record, the data in Table 3 may be extracted. For practical purposes it is assumed that 5 mm or less per month is equivalent to zero.

The length of each dry season is listed in ascending order, and the number of seasons or years in which the given duration was equaled or exceeded are counted. Dividing the number of occurrences by the length of the record, 20, gives the probability in percentage of each duration occurring in a given year. Converting to recurrence interval, we obtain the average frequency of recurrence (over a long period) of a drought of any given length. In Table 3, we find that there is a dry season of at least 3 months every year. The duration increased up to 9 months, which occurred once in the 20 years. The average length of the dry season is between 5 and 6 months. If the probability percentages are plotted on normal distribution paper it is found that between 5 and 80% of the points fall very close to a straight line, indicating that the distribution is approximately normal.

The month in which the dry season ends is also of importance in planning soil conservation measures in semiarid regions, since the most critical period for controlling erosion, as we have seen, is during the first months of the rainy season. The Kerman record indicates a relatively small range of variation in the dry season. Neglecting all rainfall of 5 mm or less, the last dry month occurred 5 times in October, 11 in November, and 3 in December. The start of the drought showed a somewhat larger variation, with two times in April, five in May, ten in June, two in July, and one in August.

C. Monthly Rainfall Distribution

Various analyses may be done on the long-term monthly rainfall values. By the conventional statistical methods, the average and standard deviation can be worked out for each month; for example, using the data from Kerman for the month of March, the average monthly precipitation, $y = 31.77$ mm, and the standard deviation, $s = 30.34$ mm. The quotient of these two values, $y/s = 0.95$, is called the coefficient of variability. This is a measure of the irregularity of the rainfall in March, and in this particular case it is quite high. By comparison it was found in a study of many stations in Ethiopia, East Africa, and Eritrea[5] that the coefficient of variability varied inversely with the average annual rainfall, and was about 0.15 for annual precipitations of 1300 to 2400 mm, and about 0.30 for annual precipitations from 500 to 800 mm.

It may be more useful, however, to calculate the probabilities of rainfall of various depths for each month or, reversing the dependent and independent variables, to calculate the depth of rainfall which can be expected at given recurrence intervals. The method is exactly the same as that used in Section II.A for the annual rainfall values. The results of this analysis are useful in estimating yields for water harvesting projects, predicting water requirements for irrigation, and assessing the likelihood of success of various rainfed crops. The latter include both those grown for economic return and those used for soil conservation purposes, such as buffer strips or grassed waterways.

III. RAINFALL INTENSITY

Rainfall intensity is defined as the rate of the fall of rain, usually expressed in inches per hour in the FPS (foot-pound-sec or English system) system, or millimeters per hour in the metric system. The intensity is rarely constant or steady during any given storm. It varies with the duration of the storm as well as with the area represented by a point reading of a rain gauge. Higher average intensities are recorded for the shorter durations and for the smaller representative areas. For any set of these conditions the intensity increases for longer recurrence intervals (or lower probabilities of occurrence). Data gathered in many places indicate that the relationship between rainfall intensity and its duration can be expressed by a formula of the following type:

$$I = \frac{KT^a}{t^b} \tag{2}$$

where I is the rainfall intensity in millimeters per hour (or in inches per hour), T is the return period of the event in years, t is the duration of the given intensity (not the duration of the entire storm), and K, a, and b are empirical constants, When I is plotted against t on log-log paper a series of parallel lines is obtained, one for each return period. The slope of the lines is b and the spacing between the lines is a function of a. K is an experimental constant which also takes into account the units used. The rainfall intensities measured over a period of 24 years in Jerusalem for example, resulted in the following expression:

$$I = \frac{15\ T^{0.2}}{t^{0.51}} \tag{3}$$

Thus, the 0.5 and the 1.0/hr intensities for the 5-year storm are 29.5 and 21.6 mm/hr, respectively. For the 20-year storm, they are 39 and 27 mm/hr, respectively.

The constants for this type of formula have been developed in several countries. In the U.S. they are plotted on maps of the country.[11] They may also be presented as graphs on log-log paper. It is interesting to compare intensities in various parts of the world. A few illustrative examples are given for the 1-hr storm with a 10-year recurrence interval:

Location	Intensity (mm/hr)
New Orleans, La.	82.6
Kano, Nigeria	81.0
Syracuse, N.Y.	38.1
Southern California	31.8
Jerusalem, Israel	24.0

There is obviously a wide spread for the recorded intensities in different locations for the same duration and return period.

To obtain a fairly complete picture of rainfall intensities in a given region, it is necessary to measure, record, and analyze readings from a number of automatic recording stations over a considerable period of time. Such data has been collected and analyzed in the U.S. and some other countries which have a long history of climatological studies and a dense network of stations. However, for most countries, particularly those of the Third World, such data is either nonexistent or very incomplete. Many of them do not have the resources necessary to establish and operate numerous rainfall intensity recorders. It is necessary, therefore, to find some criteria whereby data from other countries might be "borrowed" or adapted with a minimum of hazard.

Early studies by Bernard[3] have shown that for the entire U.S. east of the Rocky Mountains, the values of the exponent, b, for durations up to 1 hr vary within the narrow range of from 0.38 in the south to 0.47 in the north. For durations longer than 60 min, b varied from 0.78 in the south to 0.83 in the north. This is an impressive degree of uniformity over a wide geographic range, and hence could be "borrowed" for use in other regions with a certain measure of confidence.

Similarly, a, the exponent of T, varies over a narrow range from 0.15 in Louisiana to 0.23 in the region of the Great Lakes. These values might also be applied with some reliability in other parts of the world. The values of K, on the other hand, vary over a wide range in different parts of the U.S., and can be expected to differ considerably from values in other parts of the world. It is possible, therefore, to generate an approximate rainfall intensity equation for an area with limited recorded data if values of K can be determined locally from at least a few storms with 1- to 5-year return periods. With these few points, and the assumed or borrowed average values for the exponents a and b, a family of parallel lines can be extrapolated on log-log paper. Needless to say, such a graph would be far less reliable than one produced by careful analysis of long-term records, but it is a starting point for the engineer who is under pressure to produce hydrologic estimates quickly. As more field data is collected the graphs can be adjusted from time to time. The subject of rainfall intensity is discussed further in Chapter 3.

IV. RUNOFF

Runoff is defined as that portion of the rainfall which is neither absorbed into the ground, stored on the surface, nor evaporated, but which flows over the land. Estimation or prediction of runoff is of primary importance in the design of conservation measures since the flowing water is both a cause of erosion and a source of water supply. Two types of runoff will be considered in the context of semiarid zones: overland flow from small watersheds and intermittent streams. Perennial rivers in semiarid zones generally have very large catchment areas and their hydrology falls outside the scope of this volume.

Most soil conservation engineering is done on small watersheds, since one of the principles of conservation is to control the flow of water as close as possible to the place where it starts. This is sometimes referred to as "upstream engineering". In the humid regions the emphasis is on safe disposal of the runoff, i.e., in a manner which will cause minimum soil erosion. In the semiarid regions the emphasis is more upon retaining the runoff for agricultural purposes as well as for human and animal consumption. In the former case, the important characteristic of the runoff, for design purposes, is the peak rate of discharge. In the semiarid regions, peak flood flow is also important in some instances, but it is always necessary to estimate flood volumes, durations, and frequencies.

The problem of predicting surface runoff is very complicated. Literally thousands of papers have been published reporting the results of data processed according to various theoretical approaches. These vary from simplistic models relating runoff to catchment area and return period, to highly complex mathematical models which take into consideration a large number of watershed parameters, and are solvable only with the use of computers.

From the point of view of the practical design engineer there are several drawbacks to the latter approach. It requires a very good bank of basic data, climatological, hydrological, geological, agricultural, etc., which is often unavailable. The studies are time consuming and expensive, and require the services of highly trained professionals. The final results are, after a relatively large effort and expenditure, generally site specific, with limited relevance to other areas. For example, a very extensive research program was conducted on the Clarance River Valley in Australia using over 200 rainfall- and over 40 stream-gauging stations on a wide range of catchments. The mass of data collected was processed with sophisticated mathematical models. Much useful information was summarized for the basins under study, but in their general conclusions the authors stated among others:

"However, the results of the analyses . . . throw doubt on the very concepts of homogeneous hydrological regions and representative catchments. It is possible to develop relationships for transferring hydrological data from one catchment to another in a given region, but the form of these relations may vary from region to region. Unless greater understanding of fundamental hydrological processes can be obtained, it seems essential that the relationships be developed for each region from observed data on many catchments before transfer of data and results can be made with any confidence."[1]

This conclusion can be most discouraging to the field engineer, especially if he is working in a less developed region. He will heave a sigh and turn to the older literature which will give him simple, empirical rules of thumb, however incorrect they may be for his specific site. He will then try to allow for the uncertainty factor by a degree of overdesign. This process can be refined over the years by close observation of the results in the field, with corrective feedback for subsequent designs. The problem of overdesign, however, is not only one of economics; there is sometimes a question as to which are the "safe" and the "unsafe" directions of the computations. As an example from another discipline, an airplane wing subject to great stress can be overdesigned for strength, but underdesigned for safety because of the excessive weight. There are analogous situations in the design of hydraulic structures which will be discussed subsequently for specific cases.

V. PEAK DISCHARGE ON SMALL WATERSHEDS

Even in semiarid regions it is necessary to estimate the peak rates of discharge for designing spillways, culverts, waterways, and other outlet structures which will carry the outflow at the height of the rainy season when the soil storage capacity may be saturated. A time-honored method for calculating this is the so-called rational method, developed by Ramser. The formula is $Q = CIA$, where Q is the discharge in cubic meters per hour, A is the catchment area in units of 1000 square meters, I is the maximum design rainfall intensity in meters per hour, and C is the ratio of runoff to rainfall. In the FPS system, Q is in cfs (cubic feet per second), A is in acres, and I is in inches per hour.

To use this formula it is necessary to calculate or estimate the time of concentration, TC, which is the time it takes the flow to reach the point of design from the most remote point in the watershed. This is usually done by measuring the longest path of flow on a topographic map, determining the longitudinal slopes, and estimating the velocity of flow by the Manning formula if the cross section of the stream bed to the high water mark can be measured. If the stream is inspected at flood time the velocity can be estimated by timing the velocity of chips floating over a measured distance, and/or by accumulated field experience. Once the TC has been estimated, it is necessary to determine the I for a duration equal to TC. It can be shown that any rainfall intensity having a duration longer or shorter than TC will give less than the maximum Q. It is necessary to have tables or graphs of rainfall intensities for various durations and recurrence intervals. The durations relevant to small watersheds may be very short, from 5 min to about 2 hr. The smaller the watershed, the shorter the duration and the higher the intensity. As has already been pointed out, good data on the variation of rainfall intensity with very short durations is available mainly in the more hydrologically advanced countries.[11] However, in the U.S. only the Southwest and Pacific regions may be similar to other semiarid regions of the world. In Israel much detailed computation of rainfall intensities has been done[10] which may serve as guidelines for other countries with a Mediterranean type of climate. Nevertheless, borrowing intensity formulas for use in other regions must be done with appropriate caution.

Evaluation of the runoff coefficient, C, may also be problematic. Many tables of peak C have been proposed in various regions for a wide range of watershed characteristics. These are sometimes complicated by too many variables, each of which is, nonetheless, too important to ignore. It should be borne in mind that C is never really a constant, and for a given watershed it may vary as follows:

1. The annual C is the quotient of the volume of total discharge of a stream by the volume of annual rainfall which fell on the watershed above the point of control. This data is rarely available for small watersheds, but will be referred to again under the discussion of intermittent rivers.
2. The C for a single storm is usually higher than the annual C, but if there is no automatic recording gauging station on the stream, the volume of discharge will be difficult to estimate. If the watershed supplies a reservoir for which there is a depth-volume calibration, the discharge can be determined with accuracy.
3. The maximum value of C for some short period of time in the total duration of the storm would be the correction value to use in the rational formula, but this is almost impossible to determine.

In view of the many uncertainties listed above, as well as the chronic lack of reliable and comprehensive experimental data, the use of the rational formula is becoming less popular, despite its logical beauty and simplicity.

There are several alternative methods for arriving at some sort of reliable estimate of peak runoff. They are all empirical, i.e., approximate generalizations of accumulated observations. The most common form of an empirical peak runoff formula is

$$Q = K A^x T^b \tag{4}$$

in which Q is the peak discharge and A is the area of the catchment; K is a coefficient which both adjusts for the units of Q and A and allows for different watershed characteristics; T is the average recurrence period in years. The exponents x and b are experimental constants. The exponent b will be similar to that used in the rainfall intensity formulas discussed above. Exponent x has the same value in the metric and the FPS systems and is the slope of a family of straight, parallel lines obtained by plotting peak discharge, Q against area, A, on log-log paper. It has been found over a wide range of geographical conditions that this exponent, x, tends to fall within a narrow range of from 0.5 to 0.7. As in the rainfall intensity graphs, the vertical distances between the parallel lines represent various recurrence intervals. Values for K have been developed for different regions, and are sometimes borrowed for use in other areas for which the watershed characteristics seem to be similar. Once reasonable values for the constants are arrived at the use of this type of formula is extremely simple, as it is less sensitive than the rational formula, and the possible overdesign resulting from the use of K values which are too high will usually not be of great economic significance on small watersheds. An example of the experimental exponential formula will be presented later in this chapter.

VI. RUNOFF VOLUME ON SMALL WATERSHEDS

The foregoing formulas apply not only to humid zones, but also to semiarid zones when the peak flow occurs at the height of the rainy season, at which time the soil is saturated. The estimation of runoff from semiarid watersheds at other times is a different problem. The main concern in this case is the volume of runoff rather than the peak rate of discharge. Complications are caused by the erratic distribution of the rainfall and the variability of the infiltration rate at different levels of soil moisture and surface conditions. It is generally not possible to simply equate runoff volumes to total rainfall multiplied by a coefficient, because not all rainfall will produce runoff, nor will equal amounts of precipitation result in equal volumes of runoff. Small depths of precipitation falling on dry soil will be largely absorbed. What does not infiltrate may be temporarily caught in surface storage on plant surfaces and in small surface depressions and quickly evaporate.

It is necessary to observe and estimate the minimum rainfall which will produce runoff for different watershed conditions. For example, if it is noted in the Kerman area (Table 1) that runoff will generally begin only after 10 mm have fallen, the effective rainfall (in terms of producing flow) for the year 1958—1959 would include only the months of January and May with a total of 69 mm instead of the recorded 95.5 mm. Similar calculations can be made for other years. The use of monthly rainfall totals may also be misleading as the recorded rain may have fallen not in a single storm but in several small rains of less than 10 mm each. A more accurate picture can be obtained from an analysis of the daily records, if available. This involves a great deal of work which may not be justified, considering the relatively small scope of many conservation projects.

However, if the designer is close to the field and has accumulated his own experience, and that of reliable local residents, he will be able to apply sound judgment to the use of the monthly rainfall records for predicting runoff. No office study, however theoretically sound, can be considered complete without the tempering influence of field observation and experience.

Runoff as a percentage of rainfall is difficult to predict for a given storm. It is easier to work with long-term seasonal averages in which total runoff is compared with total rainfall for the hydrologic year. Recording flow gauges are rarely available on small watersheds, except on experimental plots, but it may be possible to catch the entire runoff in a reservoir and compute the volume from calibrated stage heights. This can then be related to the antecedent effective rainfall on the catchment. With a few such pond measurements distributed around the area, reasonable approximations of the annual average C may be arrived at after a few seasons. It will usually be discovered that the C calculated on a seasonal basis is lower than that calculated per single storm. Since the prediction of runoff volumes from small watersheds is required for water harvesting schemes, the conservative design procedure, allowing for uncertainties, would be to assume smaller rather than larger runoff coefficients.

VII. RUNOFF FROM INTERMITTENT STREAMS

In semiarid zones, many of the watercourses are intermittent. It is important to know their flood volumes and rates of discharge for purposes of water supply, flood control, and stream bank erosion control. The best source of data is undoubtedly the recording stream gauging station. If this is properly installed and maintained, and of sufficient length of record, probabilities of occurrence can be calculated for different stage levels, rates of discharge, durations, and volumes of flow. If recording stations exist on several streams within the same type of physiographic area, a regional pattern of stream flow characteristics may be developed which can be cautiously applied to similar streams for which no direct recorded data are available.

Table 4
TRIBUTARY CATCHMENTS TO
THE ARAVA VALLEY

No.	Name	Area km²	Record (years)
1	Zin-Scorpion	1,127	20
2	Zin-Masos	700	10
3	Zin-Avdat	233	18
4	Zin-Upper	122	16
5	Nikrot	720	14
6	Paran-Bottleneck	3,570	20
7	Paran-Halamish	2,880	10
8	Arod	160	13
9	Karkum	212	13
10	Zihor	165	11
11	Hiyum	710	11

To illustrate one type of procedure, an example is presented in some detail from a study made by the author on the Arava Valley in Israel. This is a narrow rift valley some 200-km long, lying between the southern end of the Dead Sea and the Gulf of Eilat. The stream flow is intermittent with sudden flash floods separated by long dry periods. The average annual rainfall varies from 200 mm in the north to 50 mm in the south. The value of this example to designers facing similar problems is more in the procedure followed than in the numerical results, which may be site specific.

A. Occurrence of Floods

Table 4 lists the seven tributary streams entering the Arava Valley from the west, with the catchment areas above the gauging stations and the length of record.

Over the entire period, a total of 248 floods were registered as shown in Table 5. The average number of floods per year on each stream varied from 1.23 to 2.44. The total number of floods of a given frequency was plotted against that frequency, as shown in Figure 1. It was found that these points fitted well to a Poisson distribution.*

Table 5 shows the annual volume of flood discharge (in 1000 cubic meters) for each gauging station.

The significance of the fact that the occurrence of floods on all of the catchments follows the Poisson distribution, is that each flood, on any stream, is entirely a random chance occurrence and independent of any other floods on that stream or any other stream in the area. It also indicates a certain homogeneity in the various catchments insofar as the probability of occurrence of floods is concerned. Consequently, all of the data on flood occurrences, volumes, durations, and frequencies in the Arava Valley are taken to belong to the same statistical population. This is the logical justification for the procedures to be followed in the following subsection.

B. Maximum Annual Flood

Since the catchments above the gauging stations have a relatively wide range of area,

* "A random variable, X, is said to have a Poisson Distribution if its probability distribution can be written as

$$P = \{x = k\} = P_x(k) = \frac{\lambda^k e^{-\lambda}}{K!} \tag{5}$$

where L is a positive constant (the parameter of this distribution), and K is any non-negative integer."[12]

Table 5
ANNUAL VOLUME OF FLOOD DISCHARGE (IN 1000 m³) FOR EACH GAUGING STATION

No.	Hydrologic year	Arod	Karkum	Nikrot	Zin (Upper)	Zin (Mapal)	Zin (Masos)	Zin (Scorpion)	Paran (Halamish)	Paran (Bottleneck)	Total
1	1951/52							588		2,845	
2	1952/53							630		5	
3	1953/54							1,318		51	
4	1954/55					739		806	806	51	
5	1955/56					294	362	—	—	1,505	
6	1956/57					55	379	958	—	100	
7	1957/58				77			538	—	120	
8	1958/59		10	26	76	907	1,767			45	
9	1959/60	3	20	13	44	386	491				
10	1960/61	363	50	83	51	285	566	1,646		140	
11	1961/62	2,433	924	1,814	87	252	1,139	1,075		3,875	
12	1962/63	2	0	2					366	5,208	
13	1963/64	920	319	36	654	1,176	2,128		555	(971)	
14	1964/65	396	688	235	5	218	1,178		5,514	5,418	
15	1965/66	270	0	3,708		856	969		3,000	744	
16	1966/67	1,270	470	274	42		898	1,730	2,000	4,849	
17	1967/68	84	100	213	37	470	2,479	3,477	4,455	(1,803)	
18	1968/69	1,293	1,495	282	134	672	695	958	7,760	9,811	
19	1969/70	1,226		4			174		151	520	
20	1970/71	386		487	101	722	397	1,158		20,181	
21	Total	8,646	4,076	7,177	1,308	7,032	13,622	14,882	24,607	58,242	
22	No. of floods	24	15	30	20	21	38	35	22	43	248
23	Avg. vol.	361	270	239	65	335	358	423	1,082	1,358	
24	Avg. no floods/year	2.0	1.36	2.30	1.33	1.23	2.37	1.75	2.44	2.15	1.9

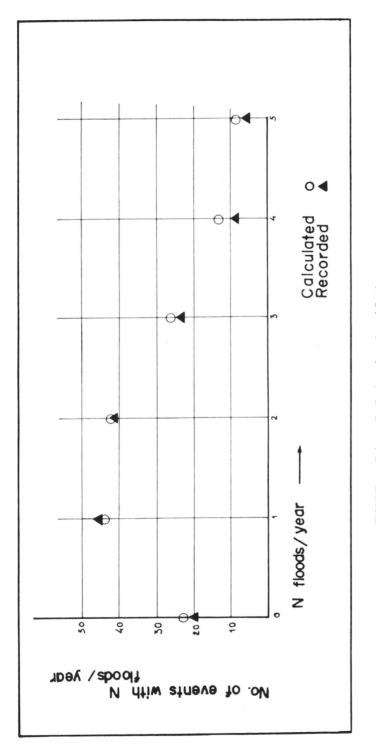

FIGURE 1. Poisson distribution of number of floods per year.

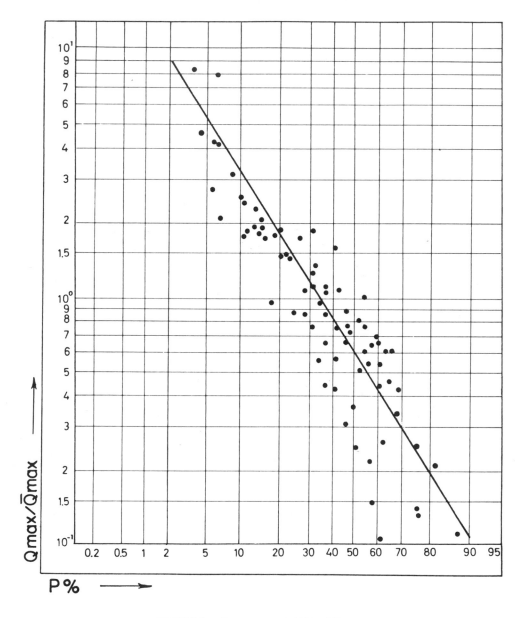

FIGURE 2.   Percentage probability of Q_r.

from 165 to 3570 km², a method was developed to generalize the data from such disparate areas. The peak annual flood, Q_{max}, is defined as the largest flow of record in each season, regardless of its size. It may even be zero in some years. For a record of n years there will be n values for Q_{max} for each station. The average of all these maxima is $\overline{Q}_{max}$. The quotient

$$\frac{Q_{max}}{\overline{Q}_{max}} = Q_r \qquad (6)$$

All values of Q_r were calculated along with their percentage probabilities of occurrence. These discharges were plotted against their probabilities on log-probability paper, as shown in Figure 2. A straight line was fitted statistically to these points, and the scatter was not great. This shows that the streams were of a similar hydrological character. The average of all the maximum annual floods for each station was calculated and plotted on log-log paper against the area of the catchment (see Figure 3). A straight line was passed through the points which represented the best fit, with a coefficient of correlation of 0.84. The function expressed by this line is

$$Q_{avg} = 0.52 \; A^{0.65} \tag{7}$$

where Q_{avg} is the average annual peak discharge in cubic meters per second and A is the catchment area in square kilometers.

It is now possible to calculate the maximum annual discharge for each station or for any catchment area by the following procedure:

1. Determine the catchment area, A, in square kilometers.
2. Read the Q_{avg} from Figure 3.
3. Select the percentage probability (or recurrence interval).
4. Read Q_r from Figure 2.
5. Calculate

$$Q_{max} = Q_r \; \times \; \overline{Q}_{max} \tag{8}$$

By this procedure many maximum discharges were calculated for each catchment area, with their respective probabilities. These values were then plotted on log-log paper (Figure 4) and a family of parallel lines, one for each probability, were passed through the appropriate points. These lines all have a slope of 0.67 which conforms to hydrologic studies made in many parts of the world, where the slope was found to range between 0.5 and 0.75. The mathematical function of these lines is given in Table 6. Extreme values of flood discharge were also plotted on this diagram and were taken to represent long-term maxima with probabilities of occurrence of less than 0.5% (or once in 200 years). This diagram is the final product of the calculation and gives the best possible estimate from the available data of the peak discharges from the various sized watersheds, with various recurrence intervals. Of course, with the addition of gauging stations, and extension of the length of the record, the predictions will become more refined.

C. Flood Volumes

Over 100 hydrographs were available on the tributaries of the Arava Valley from which both peak discharges and total volumes of the flood could be extracted, using the appropriate calibration curves. For all of these floods, Q_{max} was plotted against the flood volume, V on log-log paper (Figure 5), and the best fitting straight line was passed through the points. The coefficient of correlation was found to be 0.84 with statistically high significance. The regression followed the following function:

$$V = 16,800 \; Q_{max}.$$

where Q is in cubic meters per sec, and V is in cubic meters. Since the slope of the line on the log-log plot was 1.0, the exponent of Q was also 1. Hence, the volume of the flood varied as a linear function of the peak discharge. This is a very important finding since there are often more data available on peak flows than on flood volumes. Peak flows can be calculated from high water marks on some controlled section such as a box culvert or a straight section of channel whose slope and roughness are known. Flood volumes require the use of automatic recording instruments.

From the diagram of Figure 4, or the functions of Table 6, it is possible to prepare a table

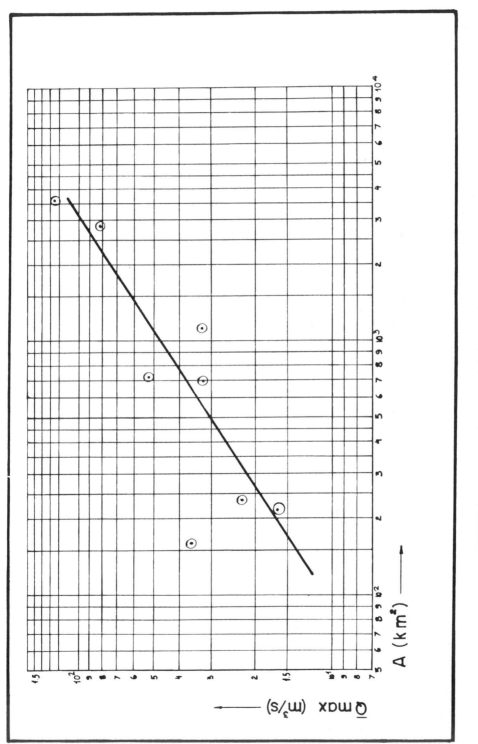

FIGURE 3.   Average maximum annual discharge against catchment area.

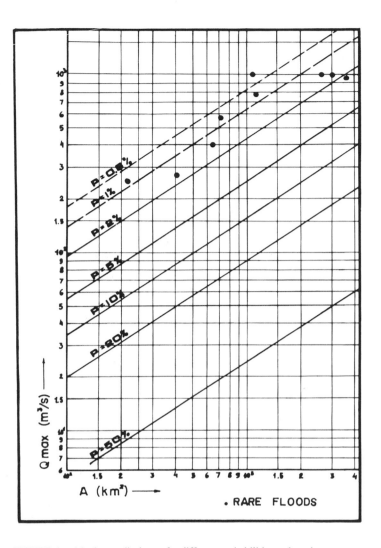

FIGURE 4.   Maximum discharge for different probabilities and catchment areas.

Table 6
MAXIMUM ANNUAL
DISCHARGES FOR
DIFFERENT
PROBABILITIES

P (%)	K[a]
50	0.24
30	0.52
20	0.86
10	1.58
5	2.81
2	4.30

[a]   In the formula, $Q_{max} = K \, A^{0.67}$,
Q in m³/sec; A in km².

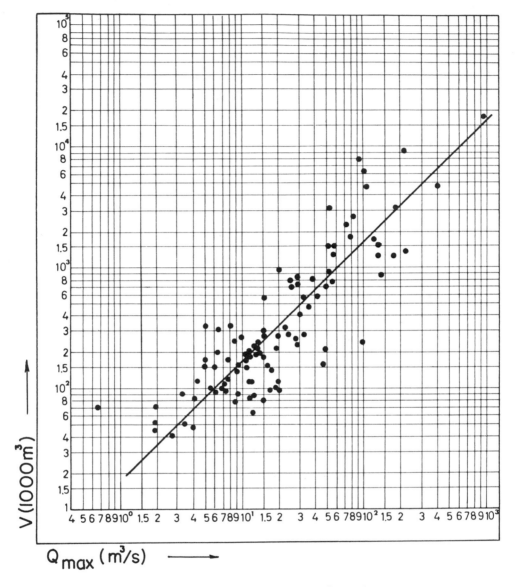

FIGURE 5.   Maximum discharge against flood volume.

of probability percentages of maximum flood volumes for watershed areas of different sizes. The exponent of A will be the same, 0.67, as it was for the formula of Q_{max}. The results are also shown in Table 7.

The recorded data on stream flows into the Arava included only a limited number of flood volumes on the larger watersheds. For floods lacking this direct measurement the volumes were calculated according to the formula

$$V = 16,800 \ Q_{max} \tag{9}$$

From the list, thus completed, of 248 flood volumes, the average volume for each station was computed and plotted on log-log paper against the size of the watershed (Figure 6). A regression line was fitted through these eight points which was found to have the function

$$V_{avg} = 8,900 \ A^{0.6} \tag{10}$$

Table 7
FORMULAS FOR PROBABILITIES OF
OCCURRENCE OF ANNUAL Q_{max} AND
ANNUAL FLOOD VOLUMES

P (%)	K in $Q = K A^{0.67}$	K in $V = K A^{0.67}$
80		0.168
50	0.24	4.03
30	0.52	8.73
20	0.86	14.45
10	1.58	26.51
5	2.81	47.21
2	4.30	72.21
1		104.1

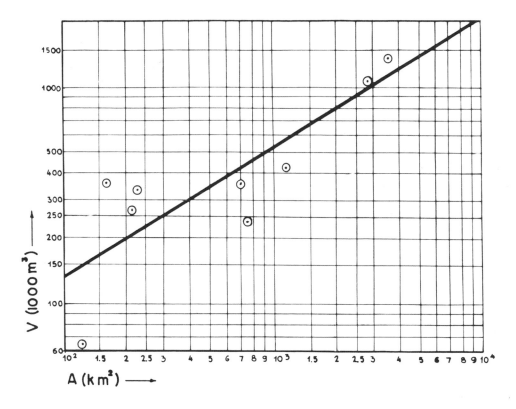

FIGURE 6.   Average flood volume for each station against catchment area.

where V is in cubic meters. Since there were an average of 1.9 floods per year, the average annual flood volume was

$$V/yr = 1.9 \times 8{,}900\ A^{0.6} = 17{,}000\ A^{0.6} \tag{11}$$

It is interesting to note that the annual flood volume is proportional to the catchment area raised to an exponent of less than 1. This indicates that the specific runoff, or runoff per unit area, decreases as the area increases. Obviously, runoff is infiltrating into the ground during flow. This is a most important characteristic of intermittent streams in arid areas. The stream channel is a *supplier* of water to the ground, rather than a *drain*, as in the case of humid regions. A sound understanding of this model will clarify many phenomena in arid zone hydrology.

D. Watershed Recharge

Since the typical Arava tributary loses water enroute, most of this loss must appear in the form of recharge to the aquifers. The methods described in this chapter can be used to estimate the average annual amount of this recharge according to the following model. A certain unit area may be considered as a "point area" from which most of the precipitation flows as runoff to a lower area. There will be little infiltration or evaporation because of the short time period involved. It is assumed that this unit area might be, perhaps, 20 km^2, representing a distance of overland flow of 5 km or less. The average yearly volume of flood runoff from this unit area would be

$$V/yr = 17,000 \ (20)^{0.6} = 102,580 \ m^3 \tag{12}$$

The total annual runoff from all the unit areas of a large catchment would be

$$V = A/20 \times 102,580 = 5,129 \ A \ (in \ m^3) \tag{13}$$

The average annual flood volume from an entire watershed is given by the formula

$$V/yr = 17,000 \ A^{0.6} \tag{14}$$

Consequently, the amount which infiltrates into the ground during the flood period will be

$$I = 5129 \ A - 17,000 \ A^{0.6} \tag{15}$$

For example, the Paran basin has an area of 3570 km^2.

$$I = 5129 \times 3570 - 17,000 \ (3570)^{0.6} = 16,000,000 \ m^3 \tag{16}$$

It is thus seen that 87% of the accumulated runoff from the unit areas infiltrated into the ground and 13% appeared as runoff at the gauging station on the Paran River. This calculation is based upon the assumption that the unit, or point area is 20 km^2, which may be arguable. Other assumptions may be tried for the application of this method to watersheds with different characteristics.

Since the overland runoff from the unit area takes a relatively short time compared to the duration of flow in the main stream, it is reasonable to assume that most of the infiltration takes place in the stream channel and not from the overland flow. This supports the concept that in the Arava type of geology, karstic limestone, the intermittent streams act as suppliers of ground water rather than as drains. It would not be true if the stream bed lay on an impervious base such as, for example, granite.

There was, however, an interesting confirmation of the method and assumptions used above. The calculation of infiltration was made for each of the tributary streams in the Arava Valley and it was found that the total infiltration, or recharge to the aquifer, in the entire west side of the valley came to an average annual volume which was very close to the recharge as estimated by the hydrogeologists using entirely different methods and data, based upon well level records, etc. The similarity between these two findings gives credence to the reliability of the model presented in this chapter.

Table 8
AVERAGE ANNUAL RUNOFF
COEFFICIENTS, C, ON TRIBUTARIES

Catchment no.	Area (km²)	Rain (mm)	V (million m³)	C (%)
1	50	45		7.9
2	20	45		11.9
3	7,218	45	3.55	1.0
4	3,540	30	2.30	2.1
5	10,758	40	5.85	1.0
6	1,554	80	1.40	1.1
7	1,704	50	1.48	1.7
8	780	30	0.89	6.1
9	984	50	1.06	2.0
10	4,092	45	2.50	1.3
11	1,116	40	1.15	2.6

Note: The names of the rivers are not given because they are only of local interest.

E. Runoff Coefficient

From the above formulas it can be shown that the average annual runoff coefficient, C, is as follows:

$$C = V/yr/1000 \; R \; A \tag{17}$$

where V/yr is the average annual volume of flood flow in cubic meters, R is the average annual rainfall in millimeters, and A is the area of the catchment in square kilometers. Therefore,

$$C = \frac{17,000 \; A^{0.6}}{1000 \; R \; A} = \frac{17}{R \; A^{0.4}} \tag{18}$$

Actual values for C on the various catchments in the Arava were calculated with the results as shown in Table 8.

It can be seen that C is larger for the smaller watersheds, and varies inversely as the 0.4 power of the area. This is a useful and practical way to arrive at a design value for C, but it should be remembered that this is an average annual value. The maximum C for an individual storm will be larger and influenced by many watershed and storm characteristics. For extremely small catchments, such as 1 km², the runoff coefficient, C, in a region of 50 mm average annual rainfall, would be 17/50 = 0.34. There is some doubt, of course, as to the permissible extrapolation to the lower limit of a micro-watershed. This should be done with caution, and corrected wherever possible by field measurements in small reservoirs, or test runoff plots. The C values obtained by the method outlined above are to be used, primarily, for calculating long-term yields in a water harvesting project.

F. Flood Duration

The duration of a flash flood is an important parameter for various engineering designs. The data available for the Arava Valley included 76 floods for which durations were measured on watersheds of several sizes. The very low flows of the hydrograph tail were neglected. The area under the hydrograph represents the volume of the flood, the length of the base is the duration, and the high point on the graph is the peak rate of discharge. For the data, a

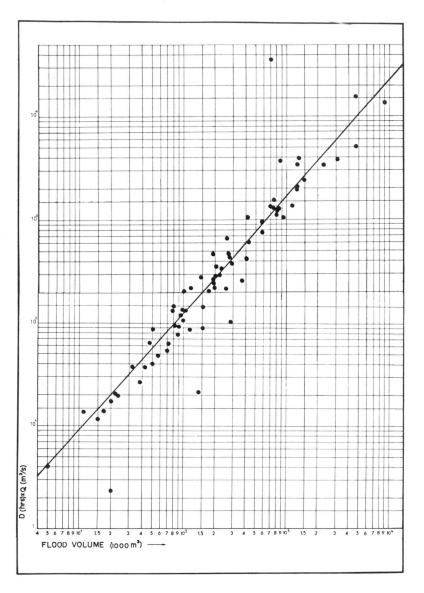

FIGURE 7.   Flood volume against duration $\times$ maximum discharge.

correlation was sought between the product of the duration and the peak flow. In such a relationship the constant of proportionality would be a factor describing the shape of the hydrograph. For each of the recorded floods the product $Q_{max} \times D$ was plotted against V on log-log paper (see Figure 7). A straight line was fitted to the points which had an extremely high correlation factor of 0.98. The function was found to be

$$D \, Q_{max} = 0.633 \, V^{1.14} \tag{19}$$

where D is the duration in hours, Q_{max} is the peak discharge in cubic meters per sec, and V is the flood volume in 1000 cubic meters. It was previously found that $V = 16,800 \, Q_{max}$. This means that the volume is a linear function of the peak flow. Substituting, we get

$$D = 15.8 \, Q_{max}^{0.14} = 10.6 \, V^{0.14} = K \, 3.657 \, A^{0.084} \tag{20}$$

where K is a constant which depends upon the probability of occurrence. It can be seen from the above equations that D is a function of either Q, V, or A, to so small an exponent that the relationship is not significant. It is concluded, therefore, that the duration of the floods in the region under study is close to a constant, varying in a range from 10 to 20 hr, regardless of the size of the watershed area. This was one of the most highly significant and useful results of the entire study. In many semiarid and arid regions, flood volumes are often not measured because of lack of automatic recording equipment at the gauging stations. Peak flows, on the other hand, are frequently determined either by direct measurement or by calculation from high water marks. The approximate duration of the flood is also known from simple observation. The flood volume can now easily be calculated by $Q_{max} \times D$ according to the above equation.

G. Summary and Conclusions

The hydrologic study of the Arava Valley was presented in some detail as an illustration of a method relevant to other similar regions of low rainfall, intermittent stream flow, and broken, hilly topography. General conclusions are the following:

1. If the average occurrence of floods from several nearby watersheds follows a Poisson distribution they may be considered as part of a single population, at least insofar as probabilities of flood occurrences are concerned.
2. Peak flows, Q_{max}, whether long-term average annual or of any given probability of occurrence, generally follow the following relationship:

$$Q_{max} = K\ A^b \tag{21}$$

    where K is a constant expressing the probability of recurrence, and b is an exponent closely approximating 0.6 to 0.7. Q-A curves on log-log paper for various probabilities will plot as a series of straight, parallel lines.
3. The relations between Q_{max} and V, flood volume, have a high linear correlation because the flood durations vary over a relatively small range, and the hydrographs have a fairly standard shape factor.
4. For watersheds of widely varying catchment areas there was a fairly constant relationship between Q_{max}/Q_{avg}, and the probabilities of occurrence.
5. The cumulative runoff from all of the "point areas" represents the bulk of the water which begins as surface flow. Most of the difference between this and the measured volume of the flood at the downstream point of measurement represents loss of water in the main channels, most of which eventually represents the amount of aquifer recharge.
6. The long-term average annual runoff coefficient for the larger catchments is small, and may range from 1 to 2%. For the so-called "point areas" it may reach ten times this figure.

REFERENCES

1. **Baron, B. C., Pilgrim, D. H., and Cordery, I.,** Hydrological relationships between small and large catchments, Austr. Water Resour. Council, Tech. Pap. No. 54, Australian Government Public Service, Canberra, Australia, 1980.
2. **Beard, L. R. and Fredrich, A. J.,** Hydrology frequency analysis, *Hydrologic Engineering Methods for Water Resources Development,* Vol. 3, The Hydrologic Engineering Center Corps of Engineers, Davis, Calif., 1975.
3. **Bernard, M.,** Modified rational method for estimating flood flows, in Low Dams, National Resources Committee, Washington, D. C., 1938.
4. **Cocheme, J. and Franquin, P.,** A Study of the Agroclimatology of the Semi-Arid Area South of the Sahara in West Africa, FAO/UNESCO/WMO Interagency Project on Agroclimatology, Food and Agriculture Organization, Rome, 1967.
5. **Cocheme, J. and Brown, L. H.,** A Study of the Agroclimatology of the Highlands of East Africa, FAO/UNESCO/WMO Interagency Project on Agroclimatology, Food and Agriculture Organization, Rome, 1969.
6. **Finkel, H. J.,** *Handbook of Irrigation Technology,* Vols. 1 and 2, CRC Press, Boca Raton, Fla., 1982.
7. **Jones, K. R. et al.,** Arid Zone Hydrology for Agricultural Development, FAO Irrigation and Drainage Pap. No. 37, Food and Agriculture Organization, Rome, 1981.
8. **Linsley, R. K. and Franzini, J.,** *Water Resources Engineering,* 2nd ed., McGraw-Hill, New York, 1972.
9. **Linsley, R. K., Kohler, M., and Paulhus, J.,** *Hydrology for Engineers,* McGraw-Hill, New York, 1973.
10. **Shein, Z. and Buras, N.,** Analysis of Rainfall Intensities in Israel, Faculty of Agricultural Engineering, Technion, Bull. No. 92, Israel Institute of Technology, Haifa, 1970 (in Hebrew).
11. U.S. Weather Bureau Rainfall Intensity-Frequency Regime, Tech. Pap. No. 29, Washington, D.C., 1958.
12. **Hillier, F. S. and Lieberman, G. J.,** *Introduction to Operations Research,* Holden-Day, San Francisco, 1967.

Chapter 3

THE SOIL EROSION PROCESS

Herman J. Finkel

I. INTRODUCTION

Soil erosion may be defined as the *detachment* and *removal* of soil material from the surface of the ground, either by water or by wind. Water erosion may take several forms such as drop splash, surface flow erosion (sometimes called sheet erosion), rill, gulley, stream bank, and channel erosion. Each of these has been amply described in the technical literature, but will be considered here with special reference to semiarid conditions. Wind erosion will be treated in a separate chapter.

Over the past 40 years in the U.S. there have been many experimental studies relating soil erosion to various parameters of soil, topography, cover, rainfall, etc. The mass of data thus collected has been summarized in a monumental study by Wischmeier and Smith,[12] and generalized into the now-celebrated Universal Soil Loss Equation (USLE). It is

$$A = R K L S C P \tag{22}$$

where A is the estimated soil loss in tons per acre per year; R is the rainfall and runoff factor expressed as the number of rainfall erosion units, corrected where necessary for snowmelt and/or irrigation; K is the soil erodibility factor, expressed as the soil loss per erosion index unit for a standard slope of 9% over a length of 72.6 ft; L is the slope length factor which is the ratio of soil loss from the field in question to that on the standard slope under identical conditions; S is the slope steepness factor which is the ratio of soil loss from the field in question to that of the standard slope under identical conditions; C is the cover and management factor which is the ratio of the soil loss under a specific management and cover to that of an area in continuous tilled fallow under otherwise identical conditions; and P is the supporting practice factor which is the ratio of soil loss under a given support practice with that of farming straight up and down the slope.*

Since the publication of this study most of the subsequent research papers on soil conservation have related their results to the USLE. This has placed all the work on a common benchmark enabling comparison of results, and the progressive improvement and refinement of the constants used in the equation. It also provides a more rational basis for the transfer of data and design criteria to other regions of the world where experimental results are less available. Frequent reference to this equation will be made in this chapter.

II. DROP SPLASH EROSION

When a drop of water falls upon the soil it breaks up into a ring of droplets which rebound from the surface in a crown-shaped structure (see Figure 1). At the moment of initial impact the kinetic energy of the falling drop detaches some of the soil particles at the surface; these rise with the droplets and are re-deposited when the droplets fall. The height to which the rebounding droplets of muddy water rise after the initial impact is surprisingly great. An illustration of this may be seen on the whitewashed wall of a building standing in a field of bare soil. The mud splash at the base of the wall may be as high as from 30 to 70 cm.

---

* To use this equation in the metric system the following conversion factors may be applied: 1 t/ha = 2.242 ton/acre, 1 t-m/ha/cm = 0.269 ton/acre-in., 1 E m = 0.683 E, 1 I (30,m) = 2.54 I (30), 1 EI (m) = 1.735 EI, 1 K (m) = 1.292 K.

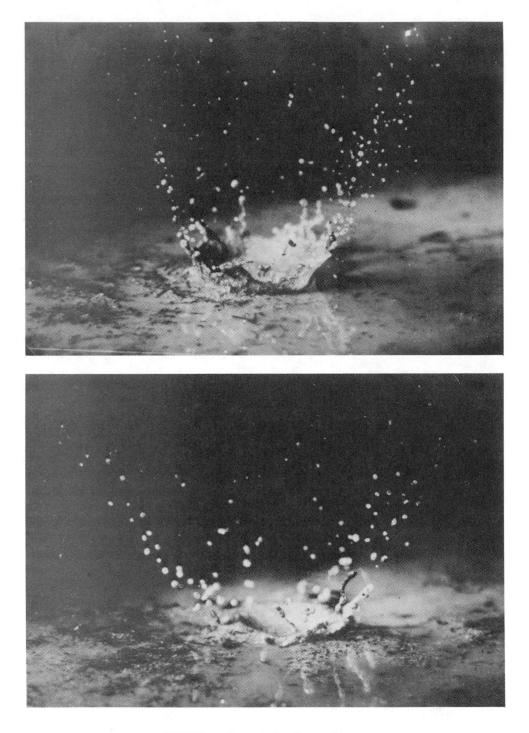

FIGURE 1.   Drop splash at instant of impact.

The horizontal distance from the point of impact, over which the muddy secondary drops may be thrown could reach as much as 2 m. On level land most of this soil falls back to the ground and little or no erosion occurs since, by definition, erosion requires both the detachment and the removal of soil from the surface of the ground.

However, when drop splash erosion occurs on a slope the trajectory of the flying droplet

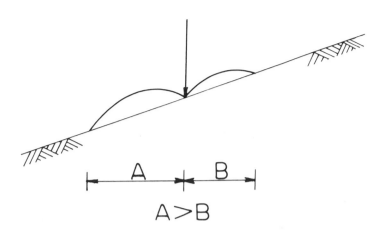

FIGURE 2. Effect of slope on distance of splash.

Table 1
RELATION OF DROP SIZE TO
PRECIPITATION INTENSITY

Precipitation intensity		Mean diameter	Mass
in./hr	mm/hr	(mm)	(mg)
0.1	2.5	1.47	1.66
0.5	12.7	2.12	4.99
1.0	23.4	2.50	8.18
2.0	50.8	2.92	13.04
3.0	76.2	3.22	17.48
4.0	101.6	3.47	21.88
5.0	127.0	3.62	24.84
6.0	154.2	3.8	28.73
7.0	177.8	3.92	31.54
8.0	203.2	4.07	35.30
9.0	228.5	4.15	37.42
10.0	254.0	4.25	40.19

strikes the ground surface at a greater distance from the point of initial impact on the downhill side than on the uphill side (see Figure 2). The net result of all drop splash is a continuous movement of soil down the slope. The amount of the displacement on the downhill side, under typical conditions, may be 10 to 15% greater than on the uphill side. With continuous rainfall over a period of hours, this represents a substantial movement of soil down the slope. The actual rate of soil erosion from drop splash depends upon a number of factors which are discussed below.

A. Characteristics of the Rain

Splash erosion is initiated by the kinetic energy of the raindrop released at the instant of impact, causing detachment of soil particles. This energy is a function of three principal characteristics of the raindrop: the size (or mass), the impact velocity, and the intensity. The size of raindrops varies up to a diameter of 7.2 mm, with the frequency curve skewed somewhat toward the larger diameters. Several studies were made of the dependence of drop size upon rainfall intensity. The average of the various findings shows that the mean diameter, D (50), which is the size above and below which the total mass of the drops is equal, is a function of the rainfall intensity, in in./hr, raised to the power of 0.23. From this function Table 1 is developed.[4]

Table 2
TERMINAL VELOCITY OF FALLING DROPS

Drop diameter (mm)	Min. fall height (m)	Terminal velocity (m/sec)
1	2.2	4.03
2	5.0	6.49
3	7.2	8.06
4	7.8	8.83
5	7.6	9.09
6	7.2	9.18

This table can be very useful in comparing the mean drop sizes from different regions. It will be found that rainfall intensities are, in general, much higher in the tropical, semi-tropical, and even the semiarid zones than they are in England and western Europe. This results in a greater erosion hazard, as will subsequently be shown.

The impact velocity is equal to the terminal velocity of the free-falling body if the drops fall from a sufficient height, which is the case with raindrops. The terminal velocity of the drops is a function of drop size. Data from studies by Laws[6] and Gunn, and Kinzer[4] are summarized in Table 2. The minimum fall height for terminal velocity listed in the table is of little importance in natural rain, but of great importance in evaluating the protection offered by various types of intercepting vegetation such as trees, agricultural crops, and grasses. It is also important in the design of sprinkler irrigation equipment which, however, falls beyond the scope of this volume.

The kinetic energy of the falling drop at the instant of impact is a measure of the power of the rain to detach soil particles. It is given by the well-known expression

$$E = \frac{1}{2} M V_t^2 \tag{23}$$

where M is the mass of the falling body and V_t is the terminal velocity. For example, in the FPS system an intensity of 1 in./hr will result in an average drop size, D (50) of 2.50 mm. This has a terminal velocity of 24 ft/sec. It can be shown that this will produce kinetic energy of about 46.5 ft·lb/ft² of land area, or 1013 ft·ton/acre. If this rain lasts for 1 hr it will be equivalent to about 1.0 hp. In the metric system, the same rain of 2.5 cm/hr will produce an energy of 685 J/m², or 6,850,000 J/ha, which is equivalent to about 700 t/m/ha. If this much rain falls in 1 hr it will develop power equal to 1.9 kW/ha. These values are, of course, only representative averages for appreciating the order of magnitude of the quantities of energy involved. In an actual rain we must separate the recorded depth into increments of varying intensities, calculate the energy delivered by each, and integrate the total.

It is interesting to compare the kinetic energy developed by an equivalent amount of water flowing over 1 acre (or 1 ha) of land during the same period of 1 hr. An intensity of 1 in./hr, with 100% runoff would give a discharge of 1 ft³/sec/acre. If this stream would flow over the entire surface at a velocity of, for instance, 1 ft/sec, the velocity head, H, would be 0.0155 ft. The power generated would be

$$P = QwH = 62.4 \times 0.0155 = 0.95 \text{ ft} \cdot \text{lb/sec, or } 0.0017 \text{ hp} \tag{24}$$

By comparison with the calculation made above the falling drops generate 576 times as much power as the flowing water. Similar calculations can be made in the metric system. Obviously, the soil detaching power of falling drops is far greater than that of a flowing stream, over a given area of land surface.

It must be borne in mind, however, that soil detachment is not necessarily equivalent to soil loss from a field. A good portion of the soil detached by splash will fall back on the same field. The actual removal of soil would be caused only by the increase of downhill splash displacement over uphill splash displacement. A greater hazard exists when drop splash (with high rate of detachment) occurs in combination with surface flow (with high rate of transport). This combination of forces leads to maximum erosion.

The assumption in the foregoing discussion that the rate of soil detachment is a function of only the kinetic energy of the raindrops is not quite correct. In the USLE, the rainfall factor, R, which is a measure of the erosive forces of rainfall and runoff, is expressed in terms of the EI, erosion index; E is the kinetic energy of the falling drops, and I is the intensity of the rainfall. It was found reasonable to posit that the total energy of a storm, E, would cause less erosion if it fell for a longer time at a low intensity than the reverse. Consequently, E is multiplied by I (30), which is the maximum intensity recorded in a period of 30 min, over a record of 22 years. Showers totaling less than 0.5 in. (or 12 mm) are not considered. It was found, from many computations all over the U.S., that the following relationship is true:

$$E = 916 + 331 \log I \tag{25}$$

where E is the kinetic energy in ft·ton/acre-in. and I is the intensity in in/hr. Values of the kinetic energy of rainfall for various intensities can be found in Wischmeier.[12*]

The EI parameters were calculated for all parts of the U.S. and presented by Wischmeier on a map of the country. In the western portion, which includes the semiarid zones, the values of the EI factor range from 20 to 50. For the eastern states the factor ranges from 75 in the north (Great Lakes Region) to 550 on the Gulf Coast. A glance at this map of isoerodent (equal EI) lines shows that it bears a great similarity to a map of isohyetal lines. It is obvious that the regions of greater annual rainfall also have higher erosiveness of the rain. However, the isoerodents are not linearly proportional to the isohyets, because, in the regions of higher total rainfall, the intensities are also higher; the intensity, it will be recalled, appears twice in the calculation of EI. Similar results were found in West Africa.[7]

In general, the semiarid regions are less susceptible to soil erosion, insofar as the precipitation factor is concerned. Work is now being conducted in many of the Third World countries to develop values for EI. Some of the studies seek to take a shortcut by establishing correlations between the EI value and some other climatological parameter, such as annual rainfall. However, whereas Wischmeier worked with a data base of 10,000 station years, many of the newly developing countries have intensity records of only a few years for a limited number of stations. Extrapolation from a very small data base is, of course, tempting, but should be done with great caution.

B. Characteristics of the Soil

Erosion results from the interaction between the erosive power of the water and the erodability of the soil. The erodability of the soil is a function of its detachability and its transportability. These properties may occur in an inverse relation to each other. Fine clay particles have a low detachability, but once separated have a high transportability. Coarse sand is exactly the opposite. The intermediate textures such as silt loam will be moderately high in both detachability and transportability and hence have the highest erodability.

This basic concept was verified by the work of Wischmeier who analyzed the results of

---

* To convert to metric units the following equivalents are given, in addition to those seen earlier: 1 t/ha = 2.43 ton/acre, 1 I(30) metric = 2.54 I (30) (FPS) (foot-pound-sec or English system) 1 EI metric = 1.735 EI (FPS) 1 K metric = 1.292 K (metric), E metric = 210 + 891 log I (metric).

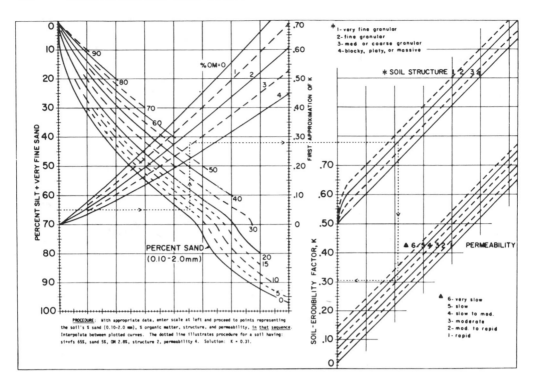

FIGURE 3.   Soil erodability nomograph.[12]

many hundreds of erosion measurement studies on different types of soil across the U.S. He found that other factors such as slope degree and length, surface treatment, and rainfall being equal, the erodability of a soil, K, varied with the parameter M, which is defined as the percentage of the silt fraction plus very fine sand. When the silt fraction does not exceed 70% the expression for K is

$$100 \; K = 2.1 \; M^{1.14} \; (12 - a) + 3.25 \; (b - 2) + 2.5 \; (c - 3) \qquad (26)$$

where a = percent organic matter, b = a soil structure code, c = profile permeability.

The determination of K is further refined by the inclusion of the element of organic matter. The greater the percentage of organic matter in the soil, the lower the erodability factor, K. Allowance is also made for the rate of infiltration. As may be expected, the soils with the higher rates of infiltration have a lower erodability. All of these factors have been combined in a graphical solution which is shown in Figure 3. It can be seen from this nomograph that the first approximation of K using only texture and organic matter does not differ greatly from the second approximation, which also uses permeability. This is because soil texture, permeability, and organic matter content are not independent characteristics but are closely related. The data base for the construction of this nomograph was sufficiently broad that it may be used with reasonable confidence in other regions of the world. However, like all broad generalizations, it should be applied with caution to any specific site and supplemented, where possible, by local measurements of erosion under controlled conditions. The main value of this soil erodability factor is in broad planning of conservation measures and the determination of priorities for areas having typical or representative soil types whose K values may be compared. Larger budgets for conservation measures should naturally be appropriated for the areas where the soil has the greater erodability.

Table 3
VALUES OF THE TOPOGRAPHIC FACTOR, LS, FOR SPECIFIC
COMBINATIONS OF SLOPE LENGTH AND STEEPNESS

Percent slope	Slope length (ft)											
	25	50	75	100	150	200	300	400	500	600	800	1000
0.2	0.060	0.069	0.075	0.080	0.086	0.092	0.099	0.105	0.110	0.114	0.121	0.126
0.5	0.073	0.083	0.090	0.096	0.104	0.110	0.119	0.126	0.132	0.137	0.145	0.152
0.8	0.086	0.098	0.107	0.113	0.123	0.130	0.141	0.149	0.156	0.162	0.171	0.179
2	0.133	0.163	0.185	0.201	0.227	0.248	0.280	0.305	0.326	0.344	0.376	0.402
3	0.190	0.233	0.264	0.287	0.325	0.354	0.400	0.437	0.466	0.492	0.536	0.573
4	0.230	0.303	0.357	0.400	0.471	0.528	0.621	0.697	0.762	0.820	0.920	1.01
5	0.268	0.379	0.464	0.536	0.656	0.758	0.928	1.07	1.20	1.31	1.52	1.69
6	0.336	0.476	0.583	0.673	0.824	0.952	1.17	1.35	1.50	1.65	1.90	2.13
8	0.496	0.701	0.859	0.992	1.21	1.41	1.72	1.98	2.22	2.43	2.81	3.24
10	0.685	0.968	1.19	1.37	1.68	1.94	2.37	2.74	3.06	3.36	3.87	4.33
12	0.903	1.28	1.56	1.80	2.21	2.55	3.13	3.61	4.04	4.42	5.11	5.71
14	1.15	1.62	1.99	2.30	2.81	3.25	3.98	4.59	5.13	5.62	6.49	7.26
16	1.42	2.01	2.46	2.84	3.48	4.01	4.92	5.68	6.35	6.95	8.03	8.98
18	1.72	2.43	2.97	3.43	4.21	3.86	5.95	6.87	7.68	8.41	9.71	10.9
20	2.04	2.88	3.53	4.08	5.00	5.77	7.07	8.16	9.12	10.0	11.5	12.9

III. THE SLOPE FACTOR

Many measurements have been made of the effect of slope factors on soil erosion. These are divided into steepness of the slope (S) and the length (L). S is related to an arbitrarily chosen uniform standard of 9%; L is related to a standard length of 72.6 ft (22 m). The reason for this arbitrary length is that the most commonly used test plot measures 6×72.6 ft, giving an area of 0.1 acre. Summarizing the hundreds of test results collected over the U.S., Wischmeier derived the following equation for the L and S factors:

$$LS = \left(\frac{\lambda}{72.6}\right)^m (65.41 \sin^2\theta + 4.56 \sin\theta + 0.065) \qquad (27)$$

where λ is the slope length in ft, θ is the angle of the slope in degrees, and m is an exponent varying with the slope, as follows:

S	M
1% or less	0.2
1 to 3%	0.3
3.5 to 4.5%	0.4
5% or more	0.5

Solutions for the above formula are given in Table 3, taken from Wischmeier.

For long, nonuniform slopes the profile may be divided into segments and the LS factor calculated for each one and added. The practical significance of the effect of S and L on erosion control will be discussed in later chapters under the appropriate conservation measures.

IV. VEGETATIVE COVER

Vegetative cover reduces splash erosion in two ways; it decreases the erosive power of the rain, and it increases the resistance of the soil surface. It is seen from the foregoing tables that a fall height of around 7 m is sufficient for all drops to reach terminal velocity. If the area is covered with a dense growth of low trees, bushes, and shrubs which intercept the falling drops, the new fall height will be considerably lower and the drops will not reach terminal velocity upon impact. Since the kinetic energy of impact is a function of the square of the velocity, the plant cover will greatly reduce the erosive power of the rain.

The quantitative effect of this interception may be illustrated by the following example: a drop with a diameter of 2.50 mm, which is the mean size, D (50), for a rain with an intensity of 1 in./hr (25 mm/hr), will have a terminal velocity of about 7 m/sec when falling from a height of 6 m or more. If the same drop falls from a height of 1 m it will have an impact velocity of 4 m/sec. From a height of 0.5 m the impact velocity will be 3 m/sec. The ratio of energy at impact will be as follows:

H	V	V^2	Ratio
7	7	49	1.00
1	4	16	0.32
0.5	3	9	0.18

In other words, the impact energy from a free falling drop of rain will be 3 times as great as the same drop falling from a bush 1-m high, and 5.5 times as great as from a vegetable crown 0.5-m high.

The resistance of the soil to erosion is also greatly enhanced by both the crowns and the roots of the vegetation. In the derivation of the USLE, Wischmeier did not attempt to separate the cover factor from that of management. The way a given crop is handled, and its stage of growth will both affect its influence on soil loss. Tables have been prepared for various types of cover and management, for different heights of growth, percentage cover of the canopy, and for the presence of mulches on the ground surface. The soil loss under these different conditions is expressed as a percentage of the soil loss for the same soil on a standard test plot tilled up and down with continuous fallow. These will not be copied here but the reader may find them in Wischmeier and Smith.[12] The practical significance of the changes in the C factor will be discussed in Chapter 4.

V. SURFACE FLOW EROSION

There are three principal ways in which soil is detached and transported by a current of water.

A. Rolling

When water flows over a soil surface particles of soil projecting into the current create increased friction. The resulting drag velocity may detach and move particles along the channel bottom. The dragged particles often move forward in short, intermittent jumps, known as saltation. As the velocity increases the movement becomes more like continuous rolling. The amount of soil moved in this way increases with the velocity of the stream and decreases with the particle size.

B. Lifting

A rough soil surface may have small clods and aggregates with pockets of air and free water between them. When a stream of water passes over such a surface it has greater horizontal velocity than the entrained water. This causes a vertical pressure differential which tends to lift the soil particles from their original positions. Once they are raised into the current they may be transported as suspensions. The rate of dislodging of the particles increases with the velocity of the stream and decreases with the internal cohesiveness of the soil.

C. Abrasion

Soil particles may be dislodged from the channel perimeter when they are struck by other, heavier particles carried in suspension. The transfer of momentum imparts a velocity to the previously stationary masses, which rise and mingle with the other suspended solids. This is called abrasion; its rate increases with the velocity of the stream and with the size and amount of abrasive material transported.

The above three processes may act simultaneously, and it is difficult to distinguish which part of the sediment load in a stream was caused by each. However, the velocity plays an important role in all of them. Any and all practices which reduce the velocity of surface flow are positive soil conservation measures of great value.

It is extremely difficult to develop an exact theoretical model for estimating the transportation of sediment by a flowing stream since, in the field, the flow is usually nonuniform, and nonsteady. The channel cross section, slope, and roughness may all vary enroute. Much empirical data has been gathered from streams by using sediment samplers. From this it appears that the sediment load which a stream can carry is a function of the velocity raised to some power between 2 and 3. Since for every velocity there is a maximum load of solids which can be transported, a stream which is close to "saturation" will not be able to carry any more material, regardless of the detachment forces in operation. Hence, a stream of clear water passing over a fairly compact soil surface has minimum abrasive capacity and maximum transport capacity. When the stream is very muddy it has maximum abrasive capacity and minimum transport capacity. Neither of these conditions will cause maximum soil erosion. The maximum will occur at intermediate values of the so-called bed load. However, if the detachment is caused by another agent, such as raindrop splash, combined with the high transport capacity of a fast and relatively clear stream, the rate of erosion will be high.

VI. TYPES OF SURFACE FLOW EROSION

Several types of erosion are caused by surface flow. They will be described briefly.

A. Sheet Erosion

In the early days of the soil conservation movement the term "sheet erosion" was applied to the condition where a more or less uniform layer of surface soil was removed from a sloping field. As this removal leaves no noticeable scars in the topography, such as rills or gullies, it is considered an insidious type of damage which, like some fatal diseases, reaches an advanced stage of development before it is recognized. It may then be too late for treatment. Sheet erosion was identified by the soil surveyors who found that on the steeper slopes with little plant cover, the so-called A-horizon of topsoil was much shallower than in the profiles of the same soil type on lesser slopes, or on the same slopes which were protected by forest or grass. The profiles with the reduced A-horizon were sometimes identified as a different soil type, which was called the "eroded phase" of the original type.

Whereas erosion undoubtedly removed a uniform depth of the upper layer, it is now felt that this is not primarily a phenomenon of surface flow, since a uniform, widespread flow of water over the surface of an entire field is a rare, if not impossible, occurrence. The slightest roughness, nonuniformity, or impediment on the soil surface will cause the flow to concentrate in the depressions to form small rills. Because of the increased hydraulic radius the velocity will increase and these minute channels will deepen. This will in turn cause the channels to erode further, and the process will continually progress. It is more reasonable to suppose, therefore, that the phenomenon known as sheet erosion is in reality the removal of surface soil by the raindrop splash process described at the beginning of this chapter. Nevertheless, being caused by an agent other than flowing water does not make sheet erosion less insidious.

B. Rill Erosion

Sloping land is sometimes observed to be dissected by a series of small, parallel channels or rills, running down the slope. Usually these do not extend back to the ridge, or watershed line, but begin some distance below it. It is useful to note this distance as a design factor for strip cropping or terrace systems. Rills usually are found on soil of high detachability. If their cross section does not increase noticeably down the slope, the factor limiting erosion may be the saturated flow which cannot transport more material. Because of roughness in the land surface the rills will eventually flow together and concentrate in fewer, deeper channels. These eventually form gullies, if the slope is long enough.

C. Gulley Erosion

The gulley is the most photogenic symbol of land damaged by erosion. It is an ugly scar in the field which cannot be crossed by farm implements, and often becomes a dumping ground for broken machinery and all sorts of rubbish. Gullies are of two main types, the V- and the U-shaped, according to their cross sections. Each of these types has a series of distinctive characteristics which indicate different measures for their control. These will be discussed in detail in a subsequent chapter.

D. Streambank Erosion

The typical intermittent stream in a semiarid region is a torrential flow, heavily laden with rocks and other sediment. As indicated in Chapter 2, it is usually of short duration with a relatively high rate of peak discharge. This causes tearing and cutting of the banks, especially on the outside of the curves. On the inside, there is a certain amount of sediment deposition. These two actions tend to increase the curvature of the stream insofar as the channel is not confined by exposed rock. The bank erosion in this type of stream is characterized by undercutting and collapsing of the wall on the outer curves. The methods of control are quite different than for stream banks of perennial rivers in the humid regions.

REFERENCES

1. **Adams, J. E.,** Influence of mulches on runoff, erosion, and soil moisture depletion, *Proc. Soil Sci. Soc. Am.,* 30, 110, 1966.
2. **Ekern, P. C.,** Problems of raindrop erosion, *Agric. Eng.,* 34, 23, 1953.
3. **Free, G. R.,** Erosion characteristics of rainfall, *Agric. Eng.,* 41, 447, 1960.
4. **Gunn, R. and Kinzer, G. D.,** The terminal velocity of fall for water droplets, *J. Meteorol.,* 6, 243, 1949.
5. **Hudson, N.,** *Soil Conservation,* Batsford Academic & Educational, London, 1981.
6. **Laws, J. O.,** Measurement of the fall velocity of raindrops, *Trans. Am. Geophys. Union,* 22, 709, 1941.
7. **Roose, E. J.,** Use of the Universal Soil Loss Equation to predict erosion in West Africa, in *Soil Erosion: Proc. Natl. Conf. on Soil Erosion,* Purdue University, W. Lafayette, Ind., 1976, 60.
8. **Rose, C. W.,** Soil detachment caused by rainfall, *Soil Sci.,* 89, 28, 1960.
9. **Seginer, I. and Morin, J.,** The Effect of Drop Impact on the Infiltration Capacity of Bare Soils, Publ. No. 62, Faculty of Agricultural Engineering Technion, Israel Institute of Technology, Haifa, 1969.
10. Soil Conservation Society of America, Proc. Natl. Conf. on Soil Erosion, Purdue University, W. Lafayette, Ind., May 24-26, 1976.
11. **Swanson, N. P., Dedrick, A. R., Weakley, H. E., and Haise, H. R.,** Evaluation of mulches for water erosion control, *Am. Soc. Agric. Eng. Trans.,* 8, 438, 1965.
12. **Wischmeier, W. H. and Smith, D. D.,** Predicting Rainfall Erosion Losses. A Guide to Conservation Planning, Handb. No. 537, U.S. Department of Agriculture, Washington, D.C., 1978.

Chapter 4

AGRONOMIC MEASURES FOR SOIL AND WATER CONSERVATION

Herman J. Finkel

I. INTRODUCTION

In prehistoric times, before mankind tilled the ground, there was a delicate ecological balance between the soil, the vegetative cover, and the rainfall. Soil gradually accreted over the centuries and the soil nutrients used by the plants were returned to the ground with plant decomposition, to be recycled for use by successive generations. Changes were gradually carved into the topography, mainly by the small geological forces operating relentlessly over thousands of years. The occasional cataclysm sometimes accelerated this process. Climatic cycles with long periods caused the advance and retreat of characteristic types of vegetation. From the point of view of the short span of a human life these changes would have been almost imperceptible.

With the human population explosion and the consequent expansion of agriculture in most parts of the temperate and tropical world this balance became seriously upset. More ground was tilled, trees and grasses were removed, and the unprotected surface was exposed to the ravages of erosion. With the disappearance of the topsoil and the exposure of the tighter subsoil, infiltration rates were reduced and the depth of moisture stored in the soil profile was lessened. This resulted in greater runoff which, in turn, led to more removal of soil. The erosion process is thus seen to be progressive with steadily growing rates of soil loss. Floods became more frequent, alternating with severe droughts. In some regions of the world this process reached the ultimate destruction, and only "badlands" remained, with gullies cut down to the bedrock.

Since the progressive process of accelerated erosion is initiated with the first removal of the natural protective vegetative cover, it is the vegetative methods of control to which we must first turn. Only when these prove inadequate should the more costly engineering, or mechanical measures be introduced. The agronomic measures are, for the most part, such that the farmer himself can put into practice.

II. LAND USE

The first and most fundamental agronomic measure for soil conservation is correct land use. The rule is that no field should be used beyond its capability to sustain a stable and permanent soil profile. Land may be classified, from the point of view of erosion, in the following scale of ascending intensity, i.e., from the lower to the higher hazard.

1. Mature woodland (unpastured)
2. Dense grassland or range (unpastured)
3. Scrub woodland (pastured)
4. Grazed pasture or rangeland
5. Close-growing crops, such as legumes and grains
6. Orchards with intertilled crops
7. Orchards with clean cultivation
8. Tilled row crops
9. Plowed fallow

This is only an approximate ranking of the erosion hazard under various types of cover. The relative positions of these land uses may change according to local conditions, cultural practices, etc. It is useful only for purposes of broad, general regional planning. On individual fields the application of various conservation practices may modify this ranking considerably.

The classical approach to the determination of correct land use for soil conservation is to develop a Land Use Capability (LUC) map based upon a survey of existing erosion. There is, however, no uniformity of approach regarding the best type of survey for this purpose. The soil survey, as a basic tool in conservation planning, has undergone considerable metamorphosis in the past 50 years. In the 1930s and 1940s great impetus was given the soil conservation movement with the rapid expansion of the U.S. Soil Conservation Service (SCS). The directors of the service at that time felt that the traditional, classical type of soil survey which had been produced slowly, county by county, was unsuitable for two reasons: (1) the emphasis was primarily on soil genesis, classification, and fertility, with specific crop recommendations; and (2) little attention was paid to the type and extent of soil erosion and suitable measures for its control. Moreover, the rate at which these surveys were produced was not adequate for the accelerated tempo of establishing conservation practices in the field. The SCS took the bold step of undertaking the necessary surveys with their own staff, in a simplified format suitable for easy use by the new army of farm planners, many of whom were agronomists, foresters, or engineers, and not professional pedologists.

The work was done in two stages. The first was a field survey of existing erosion and erosion hazards. On aerial photographs to a scale of 1:10,000 or 1:15,000 (4 in./mi in the FPS system) the obvious physical features of the land were delineated and identified by an abbreviated symbol. The main factors mapped were as follows.

Slope — Measured by a hand clinometer (Abney level) and grouped into 3 or 4 groups such as the following: (1) A = 0 to 3%, (2) B = 3 to 8%, (3) C = 8 to 15%, (4) D = over 15%. These slope groups might have different limits in different regions.

Erosion — Sheet (mild or severe); rill (average spacing); and gulley (type and average spacing).

Textural class — Determined mainly by feeling, with only occasional laboratory tests.

Depth — To B horizon or to gravel, hardpan, or other controlling layer.

Stoniness —

Special features — Added as necessary in each region, such as flooding, broken topography, etc.

Little attempt was made to identify the soil type by genesis or accepted name. A simple code for describing existing land use was added to identify cropland (without reference to specific crops), pastureland, woods, and urban areas.

When the field work was completed for a given district the areas representing each parameter were measured on the photomaps and summarized. This could be done by planimeter, but it was much easier and quicker to cut up the photos along the separation lines and collect the odd-shaped pieces for each factor in different containers. These were weighed on an accurate balance and converted to equivalent areas in the field. From this summary it was possible to know how many hectares there were in each slope group, how much area had various degrees and types of erosion, and what were the general proportions of existing land use. It was an excellent, although rough inventory of the present state of the land.

The second stage was to determine the recommended LUC for each subarea in the mapped region. This is defined as the most intensive use to which the land may be safely put, from the point of view of erosion hazard. The following classes of LUC are commonly recognized: (I) suitable for all cultivated crops, with little or no conservation measures; (II) suitable for cultivated crops, with moderate conservation measures; (III) suitable for cultivated crops, with intensive conservation measures; (IV) suitable for tree crops, with conservation measures; (V) suitable for pasture or rangeland, with controlled grazing; (VI) suitable for protected woodland; and (VII) suitable for a wildlife reserve.

The derivation of these LUC classes from the erosion survey is not automatic or unique for each combination of physical factors in the field; it is more a matter of judgment and may vary greatly from one region to another. For example, in one district a considerable portion of the land is in the A slope group with little or no erosion. This is clearly class I. There is also some land in the C slope group which could well be placed in class IV or V. In another district there is exactly the same type of land in the C slope group but little or no land in the A or B groups. Whereas technically this land should also be classified as IV or V, it would leave the district without any fields for the cultivation of crops. As this would be unacceptable to the farmers from the economic point of view, compromises must be made. It should always be borne in mind that LUC determinations must be compatible with the economic needs of the district, if they are to be followed in practice. To take an extreme case, the erosion protection of a given area may require that all of the land be retired from agriculture and revert to forest or bush. Unless the indiginous population can be transferred to another area, such a recommendation would be unrealistic.

The advantages of the early SCS system of erosion surveys and LUC classification were many. It was relatively rapid and could be carried out in the field with personnel of less professional training (a factor which may have importance even today in some of the less developed countries). It was simple to summarize, and was easy and practical to use in urgently needed conservation planning. For these and other reasons the system was copied and adapted by conservation agencies in many other countries. However, with the passage of time and the lessening of urgency to implement conservation practices it was felt that this system had some disadvantages. Many basic physical and chemical properties of the soil, as determined only by extensive laboratory tests, proved too important to ignore. These included the water holding capacity of the different layers of the profile, infiltration rates, soil structure, type of clay colloids, organic matter, pH cation exchange capacity, and availability of the various nutrient elements. Moreover, for the arid and semiarid zones several important factors were not adequately covered, such as: (1) existing irrigation and the irrigation potential; (2) salinity and alkalinity hazards; (3) redefinition of Classes IV and VI where a long dry season would limit the development of trees and shrubs without irrigation (the climax vegetation is most important to determine on land recommended to be retired from cultivation; will the semiarid bush or mesquite, which may take over, be adequate protection against erosion?); (4) in many developing countries fields are not used for a single culture alone; mixed planting of crops and intertilling of orchards are common; several species of tree crops are often grown together in the same fields; this makes clear distinctions in land use difficult to establish; and (5) shifting cultivation, which presents very special problems that will be discussed later.

Consequently, in recent years there has been a return to the more scientific soil surveys of one type or another. These take greater cognizance of the soil erosion factors as well as the possibilities of irrigation. The work is, of course, slower, but much more thorough and reliable. In many semiarid countries the mapping has made substantial progress and conservation planning is no longer greatly delayed by lack of basic soils data. Nevertheless, there are still many areas without detailed coverage and conservation agencies must either cope with the difficulties of inadequate funds and lack of personnel, or return to the quicker, but more superficial type of erosion surveys of the 1940s.

There remains the most important step of translating the survey (of whatever type) into an LUC map. This exercise must be original for each region under study. It is the art of synthesis between the physical, chemical, and biological properties of the soil, the crop requirements, the topography, the available water supply, the economics of growing each type of crop, the alternatives open to the farmer, the character and capabilities of the local rural population, the land tenure system, regional traditions and customs, and a host of other factors. It is beyond the capabilities of most soil surveyors to accomplish alone, but must be an interdisciplinary process lead by a generalist rural planner with broad experience and wide mental horizons. In addition to the various professionals on the team, the inclusion of some progressive and experienced farmers would be very helpful. To compound the difficulties, determination of the LUC for an area is not a one-time procedure. It may have to be revised from time to time according to accumulated experience, new experimental data, changing crops and market requirements, and even shifting populations.

What use can be made of the LUC maps after they are finalized? This depends upon the nature of the land tenure system and the legal force of the central authorities. If large tracts of land are held by single entities, be they private, cooperative, or governmental, it is possible to plan the land use of an extensive area according to the LUC recommendations. If, on the other hand, the land is fragmented into many small holdings which cannot be easily consolidated, agricultural planning according to the best LUC recommendations can be accomplished only by the slow process of education and extension. In some cases governmental incentives in the form of loans, grants, or subsidies can be provided to farmers who follow a correct LUC plan. Under an authoritative central regime a sound conservation plan could be carried out by decree, but the question of its maintenance over a long time is somewhat more problematical.

III. CROP SELECTION, ROTATION, AND MANAGEMENT

On the lands classified as suitable for cultivation of crops, much can be done for soil conservation by proper crop selection, rotation, and management. In many semiarid zones there are two distinct seasons, dry and rainy. The rains may come in the winter (October to April) as in the typical Mediterranean climate, or in the summer months, as in the eastern portion of South Africa. If the winter temperatures are mild enough, the main growing season will be during the rainy months, irrespective of the season. This is especially true in the lower latitudes where seasonal differences in photoperiod are minimal.

In the Mediterranean the principal rain fed crop is winter grain, which may be grown either continuously, i.e., year after year, or in some rotation. One of the main factors to be considered is the intensity of the rotation, which refers to the number of times the land is tilled and planted to a crop in the period of the rotation. The permissible intensity is usually thought to be a function of the soil fertility as only the richer soils could support a program of continuous cropping. However, the recent trend has been to support continuous cropping by heavy application of fertilizers and to ignore rotation.

Assuming that such an approach is economically justified, there are, nevertheless, limits to the response of the field to heavy chemical fertilization under semiarid conditions, without irrigation. The shortage of soil moisture limits the nutrient uptake by the plants. Furthermore, intense cultivation will lead to accelerated erosion because of the large exposure of bare, plowed land to the force of the early rains. Erosion, as has been described in the previous chapter, is a progressive process which, once initiated, grows upon itself. There are also other reasons for avoiding continuous monoculture. It is advisable to interrupt intense cropping to control plant diseases, to diminish soil compaction and aggregate breakdown which reduce infiltration rates, to improve soil microbiological activity, and to help restore the ecological balance between soil, plants, microorganisms, animal life, and climate. Apart from these reasons, the cost of chemical fertilizers in many of the Third World countries may be prohibitively high, especially in arid regions which may lie remote from the manufacturing or importing centers. Consequently, there is still a need for some form of crop rotation in many parts of the less developed, semiarid agricultural zones of the world.

The subject of crop rotations in dryland farming is given excellent treatment in Arnon,[2] a very brief summary of which follows.

Several typical rotations are practiced in the rain fed agriculture of the semiarid regions with mild winters. The predominant crop is grain, either wheat or barley. In the regions where the rainfall is between 250 to 400 mm this grain crop may be grown continuously year after year, or alternated with fallow. In some countries, such as Australia, the fallow is replaced by 1 or 2 years of a leguminous lea or pasture. This usually consists of some annual legume which selfseeds for the following year, and is grazed by cattle or sheep. In the regions with more than 400 mm of rainfall the wheat may be alternated with either weed fallow, cultivated fallow, pulses such as lentils or beans, or cultivated summer crops. It has been found that raising leguminous plants, such as pulses, for their seed, removes more nitrogen from the soil than the nodules on the roots can fix from the air. This may result in a drop of up to 30% in the yield of the succeeding wheat crop. Legumes which do not go to seed but are harvested as forage in their early stages will cause less depletion of soil fertility and moisture. The legume will also help to control weeds.

A special problem of rotation intensity arises in the semiarid regions of Africa and Latin America where shifting cultivation and bush fallow are practiced. In many villages the land is held in common and fields are assigned to each family to work for a season. This allocation is usually made by the village chief or the elders. After the field has been cropped for 2 or 3 years it is abandoned and left to bush fallow for a period which varies in different regions from 6 to 15 years. While this system is very wasteful of land it is a form of rotation suitable to the less fertile soils, where the cost of fertilizers is far beyond the means of the subsistence farmers. From the point of view of erosion control the natural cover of bush may be better than the cultivated soil, but the degree of this benefit depends upon the nature and density of the bush. A denser cover would protect the soil but it would also use up more fertility and moisture, thus defeating the purpose of shifting cultivation. A more economically useful cover to plant between successive cultivations would be some variety of leguminous pasture which can both enrich the soil and provide erosion control. It can either be harvested or grazed by animals. Such crops might be annual clovers or vetches which grow during the wet season and reseed themselves for the following year. Thus, a rotation of 1 year of winter grain (wheat or millet) followed by 3 or 4 years of leguminous pasture would be an excellent substitute for the traditional shifting cultivation. A very thorough treatment of the problems of shifting cultivation can be found in FAO[5] and Okigbo.[8]

However, there are serious obstacles to overcome. In many underdeveloped, traditional rural societies there has always been a sharp separation between animal husbandry and the growing of crops. The same people rarely work at both callings. In West Africa, for example, the division is tribal, with the nomadic Fulani (or Pehl) tribe grazing the cattle while other, settled tribes work the land. This dichotomy is as old as the strife between Cain and Abel, and is far from being bridged in modern times. In the early days of the conquest of the American West the conflict between the stockman and the farmer was sharp and often violent. It is still celebrated in ballads and folksongs. In many governmental services around the world, agriculture and animal husbandry are organized in separate ministries, often with little communication between them. The bringing together of these two fields is in itself an important measure toward soil improvement and conservation. Forage and food crops supplement each other in the rotation. The animal manure aids the growing of crops, and the grazing of crop residues is a valuable source of feed. The combining of farming and grazing is easy in modern agriculture with the use of a temporary electric fence which can be shifted around as needed. In less developed countries this solution may be unavailable or too expensive, but with the low cost of labor the grazing of plots adjacent to cultivated fields can be controlled by shepherds.

Here another obstacle arises. In many of the very underdeveloped communities of the semiarid and arid zones there is a great dearth of firewood for cooking and heating. This material is at a premium. The substitute fuel is animal dung which is collected and dried for this purpose. Although the burning of dung does deprive the soil of much needed fertility and organic matter, there has been some debate among agricultural scientists as to the true value of animal manure applied to the soil in dry, hot regions. Studies have shown that the rate at which the organic matter is oxidized by the sun is so high that much of the value is lost in a short time. If this is generally true, then the collection and drying of dung for fuel may in the long run be the most beneficial way to use it. Also, the modern equipment for converting animal manure to methane gas gives the dung a new value.

IV. TILLAGE

After rotations, the next most important factor to consider from the point of view of erosion control, is tillage. Again, taking the Mediterranean region as an example, the principal rain fed crop is the winter grain which may follow a season of either summer fallow or a summer crop which was grown on moisture accumulated from a previous winter fallow. When the wheat follows a summer crop the land will most probably be plowed for the grain in the autumn after the first rains. When it follows a summer fallow the land may be plowed either at the beginning of the summer or right after the first rains. There are advantages and disadvantages to either time of plowing, but in both cases the plowed land will be vulnerable to erosion when the stronger rains set in.

There are several ways to provide a certain degree of protection against this hazard. One is to leave a "trashy" surface on the field in the form of straw, stubble, or stover after the spring or summer crop has been harvested. Crop residues will not provide a complete protective cover, but will be far better than a clean, plowed surface. The material can be disked in before planting the new crop of winter grain. This is more easily done on the lighter soils than on the heavy. It should be borne in mind, however, that when the straw or stubble are buried in the ground their decomposition will temporarily tie up the available nitrogen to the detriment of the new seedlings. This must be offset by an application of nitrogenous fertilizer or manure unless the grain crop follows a legume and the soil is properly inoculated for the rhizobacter. On the heavier soils the disking in of a stubblefield at the end of the dry season may not provide a suitably fine seed bed for the grain, and under conventional systems of agriculture plowing would be required.

There is, however, a recent trend to question the justification of the time-honored practice of plowing. Primitive systems of agriculture do not use the moldboard type of plowshare which inverts a complete furrow slice. Lacking the heavy power machinery for this operation, the farmers in many of the so-called "backward" regions use either a shallow chisel plow, or dispense with plowing altogether and plant the new crop by hand directly in the residue of the previous one. Their results are surprisingly good. The justification of moldboard plowing was first challenged in 1942 by Faulkner in *Plowman's Folly*. The thesis of this work was that the disadvantages of plowing are many, including high cost both in money and in time, destruction of the capillarity of the soil, increase of both water and wind erosion, faster oxidation of the organic matter, development of plowsole which limits root development and decreases infiltration, and disturbance of the ecological balance of the land. This challenge to conventional agricultural science was met, for the most part, with derision on the part of the agricultural research establishment, and was dismissed as "Faulkner's Folly". Nevertheless, 40 years later Faulkner's ideas have gained a certain degree of acceptance and there are serious scientific institutions in the U.S. which now advocate the so-called "minimum tillage". Every branch of science and the arts has had, at some point in its history, challengers of the establishment who were later proved right; this does not mean, of course, that every maverick is a true prophet.

Minimum tillage may be accomplished in various ways. The land can be plowed in narrow strips coinciding with the spacing of the row crops, while leaving the intervening space untilled. The plow can be supplanted by the disk, or a type of chisel harrow, to avoid the complete inversion of the topsoil. The number of diskings and cultivations per season can be reduced. There is also a system which requires only a single trip over the field with the power machinery. The seed row is centered on the furrow slice, and a planter mechanism attached to the plow drops the seed immediately after plowing. In another method a lister throws up the soil in ridges into which the seed is planted, while the strips between the ridges are left untilled. Whereas the main objectives of minimum tillage have been to lower costs and to reduce soil compaction and aggregate breakdown, an important additional benefit is to reduce the area of plowed surface exposed to erosive rains. A fuller treatment of minimum tillage will be found in Behn[3] as well as in research reports from the Universities of Iowa, Illinois, Minnesota, Ohio, South Dakota, and Nebraska.

A more extreme position has recently been advocated called "no tillage" which recommends the complete elimination of all plowing, disking, and cultivating.[10] The basic idea is that the new crop is seeded directly into the stubble or residue of the previous crop. Weeds are controlled by herbicides so no cultivation is needed. Specially designed seeders and planters are being developed to operate in the mulchy surface. The cost of the herbicides is offset by savings in time and fuel normally spent on plowing and other tillage operations. Data is presented to show that in the U.S. Corn Belt the no-till is actually more economical than the conventional system. In addition, there is less erosion.

McGregor[7] studied the effect of the no-till system on various rotations of soybeans and corn in comparison with the conventional tillage in Mississippi where rainfall depths and intensities are high. In general, crop yields of both systems were comparable but there was a significant difference in the rate of erosion. On standard runoff plots of 0.01 ha the no-till plots over 3 years of continuous soybeans had an average annual soil loss of 2.5 ton/ha. The conventional tillage plots had 17.5 ton, or 7 times as much erosion. On the other hand, the runoff percentages from the two systems did not differ greatly, with 29% on the no-till and 23% on the tillage plots. Based upon the Wischmeier USLE, the annual soil loss under a no-till system is predicted to be 1/10 of that under tillage, for three typical soils of New York State. There is not much experimental data with respect to soil losses under no-till in semiarid conditions.

Another modern development is the so-called chemical fallow. In the western U.S., where soil moisture is a limiting factor, the land is sometimes kept in fallow for 1 year so that the accumulated moisture, of 2 years can be used to raise one crop of grain. This system, called dry farming, was normally accomplished by tilling the soil during the fallow year to remove the vegetation which would deplete the soil moisture. The tillage, however, increases the evaporation of moisture from the upper layer of the soil. It is now proposed to replace tillage fallow with chemical fallow, by keeping the weeds down with herbicides. There will also be less erosion from the first rains. The economics of this system depend upon the cost of applying the herbicides vs. the cost of the tillage. However, even if proven economical, the excessive use of chemical herbicides would be objectionable for ecological reasons and should be applied with great caution only after the possible side effects upon the environment have been carefully evaluated.

V. MULCHING

Closely related to minimum tillage is the practice of mulching. This is simply the covering of bare soil with some material which prevents or reduces the evaporation of soil moisture, while inhibiting the growth of weeds. The practice known as "stubble mulch farming" is in reality a form of minimum tillage in which the new crop is planted directly into the stubble of the previous crop. The residues are disked into the soil surface, giving rise to the temporary shortage of available nitrogen as has been previously discussed. Mulching, in the sense presently used, is somewhat different. An example is the spreading of straw over the bare

Table 1
INFILTRATION, RUNOFF,
AND EROSION UNDER
DIFFERENT MULCHES

Treatment	I (in.)	R (in.)	Erosion (t/acre)
Bare check	1.41	7.08	15.7
2 in. straw	2.10	6.39	T
1 in. gravel	3.31	5.18	T
2 in. gravel	4.86	3.63	T

Note: I = Infiltration, R = runoff, T = trace. Oat straw was 8 t/acre held with wire mesh; gravel was 0.25 to 0.5 in. diameter.

land between a row crop or between trees in a plantation. Instead of straw, other crop residues may be used, such as cornstalks, sugarbeet tops, banana leaves, etc. In some areas where crop residues are either too scarce or too valuable to use as mulch material, gravel may be used, particularly around fruit trees. The mulch reduces evaporation losses from the soil and transpiration losses from the weeds which would otherwise flourish. Mulching on sloping fields will considerably reduce the loss of soil from splash erosion by completely absorbing the energy of the impact. In addition, much time and fuel is saved by elimination of the tillage operations. The suppression of weeds is, of course, accomplished in a way that is completely harmless to the environment.

An interesting illustration of the effect of mulching is reported in Adams.[1] On an Austin Clay in Texas, with a 4% slope, there was a rainfall of 8.49 in. (216 mm) in 1 day. The proportions of infiltration and runoff, as well as the rate of erosion under different systems of mulching, were measured and are shown in Table 1.

In view of the many benefits of mulching it is surprising that it is not practiced more widely in the semiarid areas. However, it must be remembered that in the poorer farming regions of the world the many alternative uses for crop residues discourage their use as a mulch. These include animal feed, bedding in stables and barns, roof thaching, insulating material, fuel, and stuffing for mattresses and cushions.

Recommendations of soil and water conservation measures must always be weighed against the other urgent needs of the rural population.

VI. CONTOUR CULTIVATION

Years of research work and field experience have demonstrated that the cultivation of crops on the contour causes less erosion and conserves more water than cultivation up and down the slope. This is equally true in the humid and the semiarid zones, but water conservation is much more important in the latter. Nevertheless, contour farming alone, without strip cropping or terracing, has only a limited effect in reducing soil losses. It is most effective on slopes between 3 and 8%. On the lesser slopes the slope of the land approaches that of the approximate contour furrow. For slopes above 8% the capacity of the contour row to divert water is greatly lessened. At higher slopes contouring alone can actually be deleterious, since water concentrating in the furrows may break through and cause a serious washout. The same is true for the length of the slope across which contour farming is done. On a long slope the concentration of water in the contour rows may be dangerously high and the erosion resulting from a breakover would be worse than that caused by the tiny rills running down the slope.

Table 2
P VALUES FOR CONTOURING
ON VARIOUS SLOPES

		Maximum length	
Slope (%)	P	(ft)	(m)
1 to 2	0.60	400	122
3 to 5	0.50	300	91
6 to 8	0.50	200	61
9 to 12	0.60	120	37
13 to 16	0.70	80	24
17 to 20	0.80	60	18
21 to 25	0.90	50	15

The value of contouring can be expressed in terms of the P factor in the USLE (Chapter 3). This is the rate of soil loss as compared with that on a standard test plot clean tilled up and down the slope. At the SCS workshop held at Purdue University in 1956, the P values shown in Table 2 were established for contouring on slopes of various lengths and steepness.

Despite the obvious benefits of contouring, amounting to as much as 50% reduction in soil erosion, farmers often seem unwilling to follow this practice, claiming that it requires special skill, takes more time, and gives rise to a number of difficulties in tillage and crop management. In order to encourage contour farming the extension worker should be aware of these objections and know how to overcome them. The following points may be helpful.

A. Staking Out

Crop rows should be laid out as close as possible to the true contour. This is best done with an engineer's level in the hands of an experienced surveyor. As such equipment and staff are often not available to the farmer, especially in the more remote regions, simpler methods may be used. Quite accurate contour lines may be located with a hand level held on a pole with the observer sighting to another pole of equal height above the ground, held by the assistant. By moving from one marked position to the next a continuous contour line can be traced on the ground. If even a hand level is unavailable a simple T-frame or an inverted A-frame can be constructed of light lath, on the horizontal bar of which a carpenter's spirit level can be rested. By sighting along the leveled bar to a rod having a cross bar set at the same height above the ground as the horizontal bar of the A-frame, a fairly accurate contour line can be marked in the field. The line, marked by stakes will make sharp, hairpin bends across small waterways and depressions and over sharp ridges. These must be smoothed out by departures from the true contour (Figure 1). The bends across a depression can be filled with earth, if not too deep. If they are deep they should be left as uncultivated strips perpendicular to the contour lines to serve as waterways or outlet channels.

B. Parallelity

In most fields, other than those of very uniform slope, the marked contour lines will not be parallel. This creates small, triangular areas in which, if the rows are parallel to one adjacent true contour, will not be parallel to the next marked contour line. This may be handled either by a series of point rows, or with irregularly shaped correction areas which are not cultivated. These will be discussed in Section VI.

C. Plowing

The simplest type of plow for contour farming is the so-called sidehill plow which has a double set of plowshares which throw the furrow slice uphill when the plow is moving in

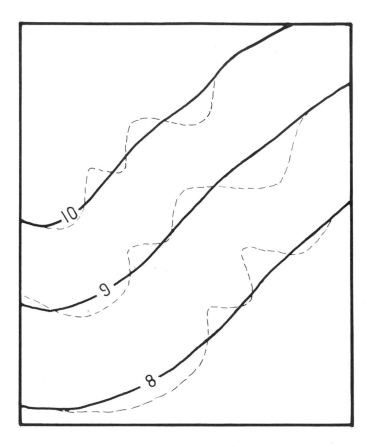

FIGURE 1.   Smoothing out the contour line.

either direction across the slope. This is accomplished by tipping the frame at the end of a row to withdraw one set of plows and insert the other for the trip in the reverse direction. The uphill throwing of the furrow slice compensates for the downward creep of the soil due to splash erosion and gravity. When a sidehill plow is not available the land must be plowed with a normal, unidirectional plow by dividing the contour area up into lands. These should not be too wide to avoid long up and down furrows at the headlands. If possible it is best to leave an unplowed turning strip at each end of the contour furrows. The land should be laid out so that the edges are on the contour, and the deadfurrow falls in the center of the land 1 year. In the succeeding year the backfurrow should fall in the center.

There is some difference of opinion regarding the importance of plowing exactly on the contour since the field, once plowed, disked, and possibly harrowed, will not have any furrows indented in the land, except for the infrequent deadfurrow between the lands. For this reason it is argued, it is quite sufficient to plow generally across the slope in straighter lines which may depart, to some extent, from the contour but which avoid many of the problems listed above. There is some justification for this point of view providing that the subsequent planting, and cultivating are done as close as possible to the true contour.

D. Planting and Cultivation

Regardless of how the field was plowed it is important that the crops be planted as close as possible to the true contour. This will determine the direction of cultivation. The various types of row cultivators all leave some form of furrow or channel which will retain rainwater if level, or which will erode into rills and gullies if it flows down the slope. Crops which

Strip	Year			
	1st	2nd	3rd	4th
A	Fallow	Wheat	Clover	Clover
B	Wheat	Clover	Clover	Fallow
C	Clover	Clover	Fallow	Wheat
D	Clover	Fallow	Wheat	Clover
E	----------------As in strip A-----------------			

FIGURE 2.   Four year rotation: Fallow-Wheat-Clover-Clover.

are ridged up or listed are more effective in retarding the runoff, but the rows must then be aligned quite accurately on the contour to prevent the damage which might occur from overtopping. With the close growing grain and forage crops there is no cultivation and it is sufficient if the rows run generally across the slope as less accuracy is required.

A note of caution must be sounded against using strict contour cultivation on heavy soils with relatively small slopes. Under such conditions it has been found that the surface drainage may be impeded to an extent which was harmful to certain crops such as potatoes. Should this condition be encountered the furrows should be made to deviate from the true contour with a longitudinal slope of 0.5 to 1.0%.

VII. CONTOUR STRIP CROPPING

If a cultivated hillside is examined for erosion it will often be found that the small rills begin to form at a certain distance from the ridge line. This distance will vary for different degrees of slope and with soil types. On the steeper slopes and for the more erodible soils the distance from the ridge to the start of the rills will be less, and vice versa. This distance can be taken as an approximation of the maximum width of contour strips into which the field may be divided. Such strips are planted to different types of vegetation with row crops alternating with close growing crops. When the close growing crop is an annual grain the cropping pattern in all the strips will shift every year. If an annual forage crop which reseeds itself is planted between the row crops the cropping pattern will shift according to the length of the rotation. An example is shown in Figure 2 for the 4-year rotation: Fallow; Wheat; Clover; Clover; For another region it might be Sorghum; Millet; Clover; Clover.

Typical widths for strip cropping range from 30 to 100 m. If wide farm machinery is used, such as seed drill, multirow planter, or combine, it is useful to make the strip widths some multiple of the widest equipment. However, the main factor in determining the width of strips is the degree of slope. Wischmeier[17] presents recommendations for strip width on different slopes and under different types of rotation, as shown in Table 3. The P values are also indicated.

If we compare these P values with those given previously for contouring we see that for alternate strips of a row crop and a small grain (type C) the values are identical. This suggests that there is no real erosion control benefit of strip cropping over simple contour farming for this rotation. For the A type rotation, however, the P values with strip cropping are just half of those for contouring alone. This, however, could be an expression of the lower intensity of the C rotation as compared with the A. The conclusion is that for better erosion control, a lower intensity rotation should be used, and this in the form of contour strips.

The lack of parallelity between the strip boundaries can be handled in several ways. In the example in Figure 3 the row crops are planted parallel to true contours marked in the middle of the strip. Their greatest deviation will be at the edges of the strip where the row crop meets the close growing crop. The close growing crop can tolerate a greater deviation from the true contour, and parallelity is also less important.

Table 3
RECOMMENDED WIDTHS OF STRIP CROPPING

Land slope (%)	P-values			Strip width		Max. slope length	
	A	**B**	**C**	**(ft)**	**(m)**	**(ft)**	**(m)**
1 to 3	0.30	0.45	0.60	130	40	800	244
3 to 5	0.25	0.38	0.50	100	30	600	183
6 to 8	0.25	0.38	0.50	100	30	400	122
9 to 12	0.30	0.45	0.60	80	24	240	73
13 to 16	0.35	0.52	0.70	80	24	160	49
17 to 20	0.40	0.60	0.80	60	18	120	37
21 to 25	0.45	0.68	0.90	50	15	100	30

Note: **A** for 4-year rotation of row crop, small grain with meadow seeding and 2 years of meadow. A second row crop can replace the small grain if meadow is established in it. **B** for 4-year rotation of 2 years row crop, winter grain with meadow seeding and 1 year meadow. **C** for alternate strips of row crop and small grain.

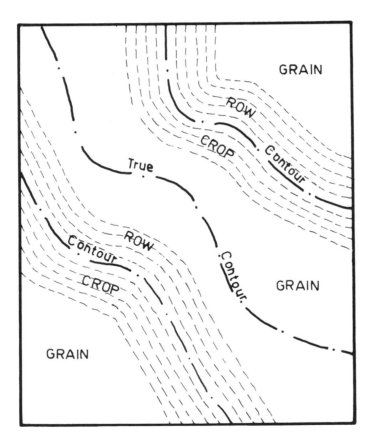

FIGURE 3.   Row crops on the true contour.

A somewhat better method is to plant both the row crop and the close growing crop on lines parallel to true contours marked in the centers of all the strips and to leave irregular areas between the strip boundaries as correction areas, sown to permanent grass (Figure 4). This method has the advantage of maximum conformity of all the strips to the true contour throughout the rotation cycle. It also leaves a permanent mark of the strip boundaries, thus

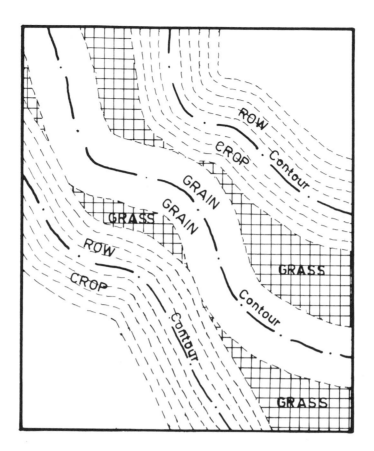

FIGURE 4. Grass correction areas in strip cropping.

eliminating the need for staking out each year. An alternative solution is to use point rows as shown in Figure 5.

Another method of making the strip boundaries permanent is to plant them with some type of low hedge, providing it will not spread and infest the cultivated fields. Such a hedgerow will also provide cover for small wildlife, such as bees, birds, etc., which are important to maintain a more balanced ecology. If the farmers object to these hedgerows as a cover for harmful pests such as rats, etc., the contour boundaries can be marked with some type of perennial native weed which is not aggressive to the adjacent crop fields. If it is bitter, such as wild onion, it will not be destroyed by grazing.

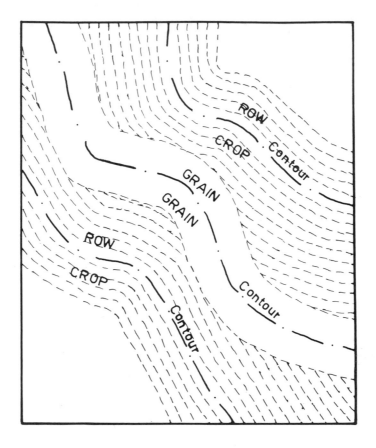

FIGURE 5.   Point rows in strip cropping.

REFERENCES

1. **Adams, J. E.,** Influence of mulches on runoff, erosion, and soil moisture depletion, *Proc. Soil Sci. Soc. Am.,* 30, 110, 1966.
2. **Arnon, I.,** *Crop Production in Dry Regions,* Vol. 1, *Background and Principles*, Vol. 2, *Systematic Treatment of the Principal Crops,* Leonard Hill, London, 1972.
3. **Behn, E. E.,** *More Profit With Less Tillage,* Wallace-Homestead, Des Moines, 1977.
4. **Brook, R. H.,** *Soil Survey Interpretation. An Annotated Bibliography,* Institute for Land Reclamation and Improvement, Wageningen, The Netherlands, 1975.
5. Food and Agriculture Organization, Guidelines: Land Evaluation for Rainfed Agriculture, Soil Bull. No. 52, FAO, Rome, 1984.
6. **Hudson, N.,** *Soil Conservation,* Batsford Academic & Educational, London, 1981.
7. **McGregor, K. C., Greer, J. D., and Gurley, G. E.,** Erosion control with no-till crop practices, *Trans. Am. Soc. Agric. Eng.,* 918, 1975.
8. **Okigbo, B. N.,** Improved Production Systems as an Alternative to Shifting Cultivation, Soils Bull. No. 53, Food and Agriculture Organization, Rome, 1984.
9. **Pereira, H. C.,** *Land Use and Water Resources in Temperate and Tropical Climates,* Cambridge University Press, Cambridge, 1973.
10. **Rice, R. W.,** *Fundamentals of No-Till Farming, American Association for Vocational and Instructional Materials,* Driftmeier Engineering Center, Athens, Ga., 1983.
11. **Roose, E. J.,** Use of the universal soil loss equation to predict erosion in West Africa, in *Proc. Natl. Conf. Soil Erosion,* Purdue University, W. Lafayette, Ind., 1976, 60.
12. Soil Conservation Society of America, Soil erosion prediction and control, *Proc. Natl. Conf. Soil Erosion,* Purdue University, W. Lafayette, Ind., 1976.
13. **Steele, J. G.,** Soil Survey Interpretation and its Use, Soils Bull., No. 8, Food and Agriculture Organization, Rome, 1967.
14. **Swanson, N. P., Dedrick, A. R., Weakley, H. E., Haise, H. R.,** Evaluation of mulches for water erosion control, *Trans. Am. Soc. Agric. Eng.,* 8, 438, 1965.
15. UNESCO, *Land Use in Semi-Arid Mediterranean Climates,* Int. Geogr. Union Symp., Iraklion, Crete, United Nations Educational, Scientific, and Cultural Organization, Paris, 1962.
16. **Walker, B. H., Ed.,** *Management of Semi-Arid Ecosystems,* Elsevier, Amsterdam, 1979.
17. **Wischmeier, W. H. and Sith, D. D.,** Predicting Rainfall Erosion Losses. A Guide to Conservation Planning, Handb. No. 573, U.S. Department of Agriculture, Washington, D. C., 1978.

Chapter 5

PASTURE AND FOREST MANAGEMENT IN THE MEDITERRANEAN UPLANDS

Ze'ev Naveh

Editor's note: Most of the chapters of this book are presented from the point of view of the engineer or the agronomist. This chapter uses the ecological approach based upon an analysis of the history and prehistory of the natural and human forces affecting the landscape, and development of a program for maximizing benefits from correct land use of pastures and woodland, consistent with the conservation and protection of the ecosystem. The Mediterranean uplands are described in considerable detail, as this is the region, above all others, where there is the longest available record of the effects of human intervention in the landscape-forming processes. It is not only typical of many semiarid regions, but may have special interest because it includes the land which is holy to many people around the world.

I. INTRODUCTION

There is probably no other region in the world that has endured more long lasting and intensive human impact throughout its history from the Pleistocene until present times than the Mediterranean Basin. There is also no other region where the unfortunate combination of a vulnerable environment and a long history of man's misuse of the land have caused such far-reaching and severe damage of soil erosion and depletion, landscape desiccation, and what is called now "desertification". Nowhere else are the dangers of combined traditional and neotechnological pressures from accelerating populations, tourists, and urban-industrial developments more threatening, causing new menaces of soil and water erosion, flooding, and land destruction.

At the same time, however, because of these long lasting and severe human pressures, nowhere else — at least in comparable climatic and ecological conditions — can the striking resilience, regenerative powers, and soil building and protective capacities of the native vegetation be demonstrated better than on the denuded Mediterranean uplands.

In this chapter an attempt will be made to point out these features and to show that they can be used as part of conservative, ecologically sound management and improvement practices for the redemption of these uplands, not only for pastoral and silvicultural uses, but also for other multiple, socioeconomic and ecological benefits, and, above all, for upland soil and water conservation.

II. ECOLOGICAL CHARACTERISTICS

A. Climate

According to the definition of semiaridness in Chapter 1, great parts of the mountainous and hilly uplands around the Mediterranean Basin, especially in the coastal foothills and lower mountains, can be characterized as semiarid. They have great erosion hazards, not only because of their topographical features, but also because of their winter rainfall regimes with typical Koeppen Cs "olive climates". These are characterized by dry, warm to hot summer months with high solar radiation and high rates of evapotranspiration, alternating with humid, cool to cold winters with lower solar radiation and low rates of evapotranspiration.

Most of these uplands have been designated in the UNESCO bioclimatic map[54] on the basis of the sum of a monthly xerothermic index (x) of biological dry days (in which total precipitation equals or is lower than half of the temperature) as "xerothermomediterranean" when 150< x >200, and "thermomediterranean accentuated" when 125< x >200. According to Naveh,[32] this is the typical "Mediterranean fire bioclimate", in which acute fire hazards prevail in the open landscape for at least 150 to 200 days.

All these semiarid uplands are well within the 400 to 800 mm isohyets. The bulk of the winter precipitation, in general, is between December and February, but there are tendencies for violent autumn and early winter rainstorms after several months of drought. These can reach, sometimes, even in the drier regions, intensities of 100 mm/hr, creating severe erosion on bare and denuded slopes.

In general, intra- and interseasonal rainfall variability is high. Thus, for instance, the relative variability of rainfall, as measured by the mean deviation from the long time average within the rainfall belt in Israel is 20 to 30%.[22] In Lower Galilee it varies from 46% in the highest rainfall month (January) to 113% in October, 77% in November, and 88% in April — the critical early and late rainfall seasons.[31] At the Neve Yaar Agricultural Experimental Station, the long-time mean annual rainfall between October and March is 552 mm. The April rainfall is neglected because it is no longer efficient for herbaceous pasture production. This annual rainfall varied within a 6-year period between 359 mm in the driest, and 690 mm in the wettest year, with a coefficient of variation of 23.3%. For December and January this parameter was 59.5%, and for February and March 49.3%. Because of these fluctuations in monthly rainfall patterns, the dry matter herbaceous pasture production was even greater and varied between 110 g/m^2 in the driest and 240 g/m^2 in the wettest year with a coefficient of variation of 65.6%. Fertilization did not only not remove these variations, but increased them even more.[37] We shall refer to these climatic fluctuations and their implication for soil and water conservation in our discussion of pasture management and improvement.

B. Physiographic Features and Soil

The physiographic and geomorphologic features of the Mediterranean uplands have been summarized recently by Bradbury.[3] They are characterized by massive young orogenic systems which evolved in the Mesozoic Era, less than 100 million years ago, but retained their maximum relief of moderate to steep slopes with high, rugged, folded, and faulted mountains and hills often rising close to the coast. Sedimentary limestones of varying hardness predominate the geological basis of the region. Weathering and geomorphologic processes, particularly those resulting from high intensity winter rains, have transformed the uplifted landscape into a mosaic of dissected valleys, narrow coastal plains, isolated lowlands, and naturally terraced hillsides.

Natural vegetation has remained only on nontillable uplands, too steep and too rocky for cultivation, and make up 50 to 80% of the total area in most Mediterranean countries. Their soils are highly complex and varying in depth, but most are shallow and heavily eroded, especially those which were terraced, cultivated, and subsequently neglected.[39] Of these, most abundant are rather fertile and well-structured terra rossa soils, derived from hard limestone and dolomite of Upper Cretaceous and Tertiary rocks, and rendzinas, derived from soft limestone with hard calcereous Nari crusts. Much poorer are the highly calcareous pale rendzinas derived from soft limestone, chalk, and Eocenic marls, which are abundant in the eastern Mediterranean, as well as noncalcareous brown soils derived from granitic rock, sandstone, and metamorphic rocks which are abundant in the western Mediterranean.

The typical red brown variety of terra rossa is high in ferric oxides, and, since the clay is of a noncohesive type, it is a permeable soil, susceptible to erosion, and easily washed off steep denuded slopes. Thus, the soil mantle is thinner on the hillsides than at the foot of the slopes, and there are many rock outcrops.

After the destruction of the woody vegetation canopy, the upper, humus layer has been lost and only skeleton lithosols have remained. However, this effect is counterbalanced by the karstic nature in which limestone weathers, especially of the Cenomanian-Turonian formations. Here, permeable dolomite and crystalline limestone is covered with a very shallow terra rossa or brown rendzina soil mantle in irregular pockets. Closer inspection reveals that vertical erosion of fine, well-structured silt and clay particles, rich in organic matter, as well as caps of soft chalk and marl lenses, ensure favorable fertility and moisture conditions for the deeply penetrating shrub roots.

C. Vegetation

The Mediterranean uplands are part of the so-called Sclerophyll Forest Zone, in which broadleaved and chiefly evergreen trees and shrubs with thick but mostly leathery leaves (sclerophylls) reach their optimum development and distribution. Their closest ecological counterparts outside the Mediterranean are the broadleaved sclerophyll shrublands and woodlands in central and southern California and in similar bioclimatic regions in central Chile.

Di Castri[9] has recently summarized the major ecological characteristics of Mediterranean shrublands and Thirgood[51] has described those of Mediterranean forests. With the exception of the rapidly vanishing coastal dunes and undrained wetlands and marshes, these noncultivable upland ecosystems are the last refuges of natural (spontaneously) occurring and reproducing Mediterranean floras and faunas. Wherever they have not yet been converted into dense pine or eucalyptus forests or depleted into manmade scrub or asphodel and rock deserts, they are distinguished by their great biological and ecological diversity in space and in time. This has been chiefly the result of the great macro- and microsite heterogeneity of a rugged and rocky terrain and the long history of fire and human modification, lasting for hundreds of thousands of years.

These induced complex and dynamic regeneration and degradation patterns, ranging, according to site conditions and past and present land use pressures, from rich, productive, open grasslands and woodlands to severely depleted dwarf shrublands (called batha or phrygana) and manmade rock deserts, and from rich multilayered semiopen shrublands and forests to one- to two-layered, closed, tall shrublands (called maquis or mattoral).

The natural potential vegetation (in the Mediterranean still misnamed as ''climax'') of the cooler and wetter accentuated thermomediterranean bioclimatic zone consists of sclerophyll maquis shrublands. These are dominated by *Quercus calliprinos* (Kermes oak) in the eastern Mediterranean and by *Q. coccifera* in the western Mediterranean. Here, in the most favorable conditions, well developed sclerophyll forests can also be found, dominated by *Q. ilex*.

Most of these sclerophyll trees and shrubs are distinguished by dual root systems both spreading horizontally and penetrating deep into rock cracks, and by resprouting after fire, grazing, or cutting; they also respond favorably to pruning the coppicing of one stem. If resprouting from suckers is prevented by recutting or browsing, they soon attain the stature of small trees. Thereby, closed, unilayered, very fire prone, and unproductive shrub thickets can be converted into rich, multilayered park-like groves and woodlands. This, apparently, was the way sacred oak groves were created in cemeteries, which have mistakenly been regarded as remnants of ''climax oak communities''.

In the coastal foothills and lower mountain elevations in the slightly drier and warmer xerothermomediterranean region the natural potential vegetation has a park-like nature, dominated by scattered *Ceratonia siliqua*, the highly valuable carob tree, or by *Olivea europea*, the native olive tree, with a rich shrub and grass understory, dominated by *Pistacia lentiscus*, the Eumediterranean ''mastic''. This evergreen shrub, combining low palatability and great drought resistance with vigorous regeneration powers after fire and cutting, is one of the last shrubs to survive over large areas of overgrazed, depleted, mosaic-like shrub grassland with low or nonpalatable early maturing annuals. However, if these pressures are released, a striking vegetative recovery of woody plants occurs, together with a very species rich productive grass cover.

Another important formation in drier parts, consists of deciduous oak woodlands of *Quercus ithaburense* in Israel and *Q. macrolepi* in Turkey, very much resembling the Blue oak (*Q. douglasii*) woodlands in California. Here, a very species rich, productive grass and legume understory can be maintained under moderate grazing pressures. However, by overgrazing, too light grazing, or full protection, species diversity is reduced, and tall, aggressive grasses, such as *Avena sterilis* and perennial thistles, take over.[40]

Most productive herbaceous, natural pastures can be found in even drier submediterranean conditions on fertile rendzinas and basaltic soils, rich in phosphate, and therefore conducive to vigorous growth of many valuable pasture plants. The poorer, highly calcareous pale rendzinas, as well as the poor brown soils from granitic rocks, sandstone, and metamorphic rocks are much lower in fertility and highly erodible. On these, the herbaceous plants are dominated by xerophytic phanerophytes and chamephytes, including many unpalatable, aromatic Labiatae, and also *Cistus* species. Being favored by frequent burning and heavy grazing, these occupy large areas in many Mediterranean countries. Their conservation into more valuable ecosystems should be considered, therefore, as a major management challenge.

III. MAN-INDUCED MODIFICATIONS OF MEDITERRANEAN LANDSCAPES AND THEIR EFFECT ON SOIL AND WATER EROSION

A wealth of accounts has been published in the last 50 years on the history of Mediterranean lands, their uses and abuses, and their effects on the landscape and on soil and water erosion. Here, only some of the most recent reviews of these problems will be mentioned, and a short summary of their major findings, as related to upland soil and water conservation, will be presented.

The sad history of the resource depletion of Mediterranean forests has been described lucidly by Thirgood.[51] The impact of man and livestock has been reviewed in a comprehensive way by Le Houerou,[23] and that of man and fire by Trabaut.[53] Naveh[33,34] and Naveh and Dan[39] discussed the degradation and reclamation of Mediterranean uplands and their ecological management. More recently, Naveh and Lieberman[47] published a detailed account of these subjects and proposed an integrated, interdisciplinary program of landscape ecological oriented education, planning, management, and research, aimed at the reconciliation between the needs for socioeconomic advancement and the conservation of natural resources, including soil, water, plants, and wildlife. The history of these modifications can be roughly subdivided into three major phases.

A. The Evolution of the Seminatural Mediterranean Landscape of the Paleolithic Hunter-Gatherer in the Pleistocene

In this "coevolution", as described in more detail for Mt. Carmel in Israel,[38] fire has played an important role as the first extrasomatal source of energy, not only for heating and cooking, but also for food collecting and hunting through the opening of dense forests and shrublands, and the creation of more accessible and richer ecotones for man and his game.

Recent archeological findings in the Petralona limestone cave in northern Greece have provided convincing proof that the use of fire by Acheulean hunter-gatherers can be dated back for about 1 million years and can, therefore, now be considered as the oldest existing proof of fire culture in the world.[46] This early use of fire coincided most probably with the widespread occurrence of wildfires caused by volcanic activity, and by lightning and thunderstorms which apparently accompanied the establishment of Mediterranean rainfall patterns during the drier interpluvial periods in the Middle and upper Pleistocene. As reviewed by Perles,[45] this is documented by many findings from later Acheulian and Mousterian cultures in southern France, Greece, Spain, and Israel.

The sophistication in the use of fire as a major vegetation management tool, and in combination with the intensive preagricultural utilization of plant and animal resources, apparently reached a peak at the end of the last glaciation, 10,000 to 15,000 B.C. At this time, mild winter rain climates favored the spreading of epipaleolithic food collecting-hunting-fishing *Homo sapiens* populations, such as the Natufians of Mt. Carmel.[18]

It can be surmised that this coevolution led to the establishment of a dynamic flow equilibrium between man, the fire modified landscape, and its biotic resources of woody and herbaceous food and fodder plants and game animals of the intensively utilized semi-natural landscape. Adopting Jenny's[20] functional-factorial approach in which the vegetation and soil features (including erosive and hydrological processes) are dependent variables of ecosystem state factors, we can describe this flow equilibrium in a semiformal way as the first anthropogenic biofunction equation, in which not only natural geo- and biofactors, but also man through land use, induced changes in the initial site conditions:

$$E_{s,v,a} = f(H_{bu,hu\text{-}ga}, P,R,C_{dr,fi},O_{gr}...T < 100,000)$$

In this equation, s,v,a are the dependent variables of soil, vegetation, and human artifacts of the landscape unit or ecotope (E), as a function of the initial site conditions of soil parent material (P), the relief (R), and the driving flux potentials of climate (C) and organisms (O), along with their most important evolutionary forces; namely drought (dr), fire (f), and grazing (gr), operating together with the human use factor (H) of burning (bu), hunting and food gathering (hu-ga), and other, nondefined factors (. . .), during geological times T<100,000.

From this equation it is apparent that soil and water erosion, as soil vegetation variables are depending on the controlling ecotope state factors, and should therefore be treated as multifactorial processes in this equation. We have no way to find out if, at all, and how great was the effect of human interactions on erosive processes during this dynamic orogenic and morphotectonic geological period. Judging from similar flow equilibrium states which existed until recently in "primitive" preagricultural burning, hunting, and food collecting economies in comparable Mediterranean ecological conditions, such as the Coastal Indians of southern and central California[24] and the aboriginies of southwestern Australia,[17] we can assume that there were very strong adaptive, ecological, and behavioral negative feedbacks which ensured long term metastability in these human ecosystems and prevented overuse of their natural resources.

B. The Evolution of the Early Agricultural Mediterranean Landscape in the Early Holocene

The transition from intensive food collection to food production and domestication of plants and animals — the so-called neolithic revolution — was actually a gradual process of cultural evolution, lasting several thousand years in the postglacial Holocene.

As described in detail for Mt. Carmel,[38] the intentional use of fire also played an important role in this process, and especially in slash burn rotations on which the cultivation of the fertile forest soils of the lowlands was based. This practice was repeated several thousand years later by the first neolithic farmers of Europe.[29] As shown by Iversen,[19] they cleared their oak forests by chopping the tree trunks with stone axes, and used the ashes of the burned brush wood as fertilizers for their crops.[29] In the Mediterranean proper this is supported by the palynological findings of Banlieu.[2] In these agricultural systems the lower parts of the huge stamps and their extensive roots remained intact and could protect these deep vertisols from erosion.

However, after narrowing the broad spectrum of neolithic agriculture into a cereal field, crop and pastoral livestock husbandry, the complete destruction of the pristine vegetation in the lowlands and along river beds was initiated, followed by severe soil erosion. According to Dan and Yaalon,[7] this can still be witnessed in the erosive exposure of ancient, calcified paleolosols in the coastal plain of Israel.

Thus, the rise of early pastoral, rural, and urban civilizations replaced the unique mutual adaptation of physical, biological, and cultural features of the coevolution of paleolithic Mediterranean man and his seminatural landscapes by the unilateral human dominance which henceforth governs the fate of Mediterranean landscapes for better or worse.

C. The Final Formation of the Mediterranean Agro-Pastoral Landscape in Historical Times

The final formation of the Mediterranean agro-pastoral upland landscape was initiated by the domestication of fruit trees in the Bronze Age, around 5000 B.C.; however, it was completed only after the invention of iron tools, several thousand years later. These enabled the uprooting of shrubs and trees on the slopes and the clearing and terracing of upland fields. It can be assumed that wherever these slopes were too steep or rocky, favorable microsites between rock outcrops were also planted with fruit trees, such as olives, figs, almonds, pomegranates, and grapes. At the same time, however, the natural woody vegetation was apparently left for soil protection along terrace walls and on steeper slopes between the terraces. These, and the scattered rock site plantations, were protected from grazing animals by thorn or stone fences and in this way an efficient intensive, multiple, but at the same time, also conservative use could be made of the rough terrain.

These "polyrock cultures" were exported by the Phoenicians from the Levant to all over the Mediterranean Basin. They were brought to their highest level of agro- and hydrotechnical sophistication during Roman times, in combination with irrigation and water conservation, rotation of crops, manuring, and stubble burning for mineral fertilization of the fields. These intensive upland use patterns are well documented in *The Bible,* the *Talmud* and in classical literature. They can be regarded as the first instance, and, alas, one of the very few, in which agriculture improved the initial ecosystem factors of topography, soil parent material, and moisture regime on a long-term basis.

That this system was not only ecologically but also economically viable and profitable can be judged from the fact that in the 1st century, according to Flavius Josephus, the rocky Galilee mountains alone maintained a dense population of 2.5 million people, chiefly farmers, enjoying high incomes from olive oil and wine export.

However, it should also be realized that at the same time, the Mediterranean herdsman continued the practices of his prehistoric predecessor of intentional brush burning to increase the herbaceous fodder from his shrubland pastures. In addition, the steadily growing demands of these expanding and wealthy populations for timber, fuel, and charcoal led to extensive forest clearings. It was this unfortunate combination of tree and woodcutting, fire, and grazing which caused the gradual landscape desiccation, especially in the drier regions and on steeper and poorer slopes.

According to Heichelheim,[16] despite these extensive forest clearances during classical times, there seems to have been no substantial loss of soil fertility and there appears to be little evidence to support the theory of Jacks and Whirte that ascribed the decline of the Roman Empire to deforestation, soil exhaustion, and erosion. As described in detail by Thirgood,[51] there were great differences, in this respect, between different countries and even between different adjacent sites, according to their ecological conditions and the land use practices of their owners.

We should therefore be aware of the dangers of sweeping generalizations, such as the wholesale condemnation of fire, goat grazing, and woodcutting on Mediterranean uplands which have allegedly turned all once forest-covered mountain slopes into rock and scrub deserts. Most of these shallow and rocky slopes which were never cultivated nor terraced have probably not undergone any extreme changes in their vegetation since postglacial and prehistoric times, and, as shown by Naveh and Dan[39] in the example of Israel, their fertile and fine structured brown rendzina and terra rossa soils have suffered much less from erosion than has been generally assumed. This is true only if their vegetation mantle has not been uprooted and their soil has not been disturbed by cultivation or otherwise. Due to the great resilience and recuperative powers acquired during their long evolutionary history of continued pyric, ungulate, and human pressures, this hardy vegetation can provide very efficient soil protection, as long as the woody plants can regenerate vegetatively from their extensive rootstock, and the herbaceous perennials from their underground bulbs, rhizomes, and other regenerative tissues, and the annuals can draw from sufficient seed reserves in the soil and close to its surface.

On the other hand, catastrophic soil erosion, flooding, and siltation, leading to badlands and swamp formation, occurred as the result of the abandonment and neglect of the terrace walls and the disintegration of their local base levels in periods of political upheaval, warfare, and depopulation. This was the case during the decline of the Roman Empire and again after the downfall of the Byzantine Empire.

The general agricultural decline during the last 1300 years in the Levant, which has been described so vividly by Reifenberg,[47] led to the replacement of settled agriculture by pastoral nomadism. According to Taylor,[50] the depth of silt and gravel over Roman and Byzantine bridges indicates that between 2 and 4 billion m^3 of soil were washed off the western side of the central Palestinian hill country, which was mostly terraced. This would have been sufficient to form 4000 to 8000 km^2 of farming land. For other Mediterranean countries, Vita Finzi[55] has provided proof of erosive changes in river channels upstream and aggradation and sedimentation downstream since Roman times.

We may conclude by stating that during the Holocene, the Mediterranean agro-pastoralists gained complete dominance of their total human ecosystem by changing not only the dependent ecotope variables of soil and vegetation but also all other controlling state factors, with the exception of terracing, mostly for the worst. This can be expressed by the state factor equation:

$$Es,s,a = fHag-pa (P,R,C,O.....T <100)$$

with human agro-pastoral land use (Hag-pa) as the dominating state factor. During historical times, these multifactorial anthropogenic functions became a series of mandriven degradation and regeneration cycles, each lasting for several hundred years, according to the duration of the historical land use period, and therefore T<100.

IV. THE DYNAMIC AGRO-PASTORAL FLOW EQUILIBRIUM AND ITS PRESENT DISTORTION

Throughout the long history of agro-pastoral utilization, the seminatural and pastoral ecotopes of open forests, shrublands, woodlands, and grasslands, and the agricultural ecotopes of terrace patches and hand-cultivated rock polycultures became a closely interwoven dynamic cultural landscape mosaic. The transfer of fertility (through grazing) and of seeds (through grazing, wild herbivores, and insects) created ideal conditions for introgression and spontaneous hybridization of wild and cultivated plants and biotypes, and for the evolution of genotypes with high adaptation to this man-modified habitat.

Simultaneously, great variations in space and time (namely microsite and floristic diversity), and the short-term, mostly cyclic, climatically induced seasonal and annual fluctuations, contributed to the global stability and persistence of these ecotopes, even under heavy human pressures. In the traditional Mediterranean pastoral systems, great seasonal and annual fluctuations in productivity[37] apparently prevented overgrazing. The numbers of livestock which could be supported during critical periods of low food availability in early winter and in drought years were not sufficient to overgraze pastures during the spring flush of growth

FIGURE 1.   North-facing slope in the Judean Mountains.

and seed setting and, in wet years, in a way similar to self-regulating natural wildlife populations. At the same time, overcutting and overburning was also prevented by burning and coppicing rotations necessary to ensure sustained productivity and sufficient recovery.

These regular grazing, burning, and coppicing regimes led to the establishment of a dynamic equilibrium in those upland ecotopes which were neither overgrazed or overcoppiced for prolonged periods. According to Naveh and Dan,[39] this man-maintained dynamic flow equilibrium between trees, shrubs, herbs, grasses, and geophytes contributed much to the striking biological diversity and attractiveness of the open Mediterranean landscapes and became their most important asset for recreation and tourism.[9]

This human-maintained equilibrium is now disrupted by the combined and even synergistic impacts of intensified traditional and modern agro-pastoral pressures, neotechnological landscape degradation, despoilation and pollution, urban-industrial sprawl and uncontrolled mass recreation and tourism.[1,47]

Not only have the intensification of these human perturbations, but also their complete cessation in some areas, either by abandonment as a result of depopulation or by intentional protection, have disrupted this flow equilibrium by reducing ecological diversity, vitality, stability, and attractiveness.[40] Ironically, such intentional protection has been recommended by leading phytosociologists as the most desirable conservation strategy of maquis and sclerophyllous forests in order ''to reconstruct their hypothetical climax''.[52]

Naveh and Dan[39] have provided some typical examples of erosive degradation functions on the two major soil types of the semiarid Mediterranean uplands in Israel. In Figure 1, two opposing slopes are represented from the southern Judean mountains in the Beth Govrin region. The northern slope is covered with Nari — an ancient calcrete crust rock — with many outcrops, a noncalcareous brown rendzina soil, and a fairly dense canopy of maquis shrubs, dominated by *Quercus calliprinos, Phillyrea media, Pistacia palaestina* and *Pistacia lentiscus*. This slope, because of its rockiness, has apparently never been cultivated and lacks all remnants of terraces.

FIGURE 2.   South-facing slope in the Judean Mountains.

Under the present moderate grazing regime practiced by the nearby collective settlement, the slope seems to be well stabilized. On the south-facing slope, however, where the Nari has been stripped off and the underlying soft chalk has been exposed, and a highly calcareous pale rendzina has been formed,[6] the degradation pattern took a very different course. In the past this slope — like most, if not all, other pale rendzina slopes — was cut to transform the circular shape of the hill into a series of rather broad, flat, and leveled terraces, bordered by stone wall.[48] Later on, these terraces were neglected and since then, have been badly eroded. After their abandonment they were covered by dwarf shrubs and grasslands (Figure 2).

A typical profile diagram of the soil-vegetation degradation pattern on such a terrace is presented in Figure 3. Where erosion has been most severe, above the ancient terrace wall, the soft chalk has been exposed and covered with a scarce vegetation chiefly of *Thymus capitatus*. This is one of the most prominent dwarf shrubs of batha and is successful invader or erosion pavements, rock outcrops, and highly calcareous rendzinas, where heavy competition of quicker growing therophytes is prevented due to the low soil fertility level.

On the slightly less eroded belt, toward the central part of the terrace, plant cover is more dense, and *Asphodelus microcarpus* dominates. This is a tall, unpalatable geophyte with a very wide distribution, especially in heavily grazed and degraded open herbaceous communities. On the least eroded and flattest belt, beneath the terrace wall, the rather deep soil includes a definite A and C horizon.

The dense herbaceous sward is dominated by *A. microcarpus,* as well as by *Hyparrhenia hirta* and *Andropogon distachyus,* two drought-resistant perennial bunch grasses, prospering on sunny and rocky slopes. These are accompanied by *Avena sterilis,* the most abundant annual grass under grazing conditions, flowering geophytes such as *Cyclamen persicum, Anemone coronaria, Iris palaestina,* and a highly variable mixture of other annual grasses, legumes, and herbs.

On the colluvial and gravelly remnants of the broken down terrace wall, dwarf shrubs can be found, while on the stone remnants and the chalky rocks shrubs such as *Pistacia lentiscus* and *Rhamnus palaestina* can be seen.

With complete cessation of human interference, as implemented in a nearby nature reserve, the maquis shrub vegetation has been transformed into an almost impenetrable thicket, 2- to 3-m high, uniform in structure, and devoid of any herbaceous understory. A similar trend has also been observed in completely protected maquis reserves in northern Israel.

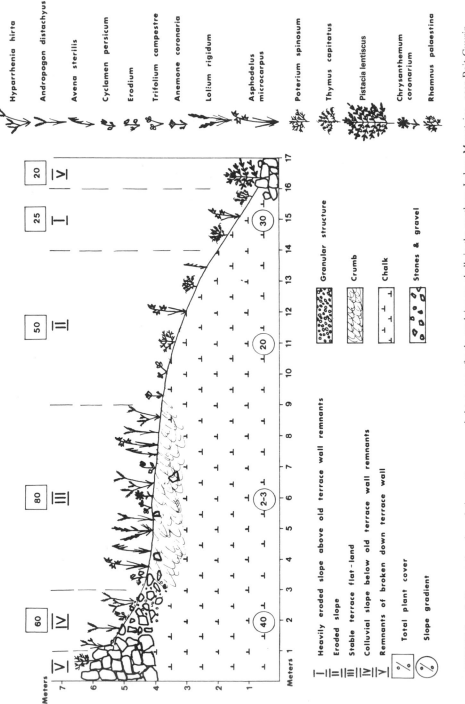

FIGURE 3.   Soil-vegetation degradation pattern on eroded terraces of pale rendzina on chalk in the southern Judean Mountains, near Beit Guvrin.

On the other hand, in the densely populated Hebron Mountains, the degradation cycle has been continued and even accelerated as a result of heavy human and livestock pressure. The pale rendzinas which had already suffered in the past from severe erosion are now cultivated again, without reclaiming or reconstructing the broken-down terrace walls. Thus, their erosion cycle has been renewed. The uneven and poor stands of the barley crop reflects the different erosion belts on these terraces.

On the Nari slopes also, small soil patches between the rock outcrops have now been plowed and sown with barley. Most of the shrubs have been cut and handgrubbed for fuel, but a dense grass cover still remains near rock edges and around the sown patches. This cover is dominated by *Poa bulbosa* and *Hordeum bulbosum,* which are not only resistant to grazing pressure and drought, but serve as effective soil binders and protectors.

The steeply inclined slopes of hard limestone and dolomite in the Hebron Mountains which have never been cultivated, and those with leveled geological bedding whose terraces were abandoned in the past,[48] are now also exposed to heavy human and livestock pressure. This pressure, superimposed on the low soil fertility and the harsh climatical conditions, has produced some of the most advanced degradation stages in Israel. They consist of shallow and eroded terra rossa and xeric brown rendzina with a sparse dwarf shrub cover (mainly *Sarcopoterium spinosum* on the northern slopes), and of bare soil with stunted herbs and geophytes (mainly *Asphodelus microcarpus* on the more exposed and drier sites).

Also in the slightly higher rainfall regions in Israel and elsewhere, as long as the Nari and karstic slopes which were never terraced and cultivated are covered by a more or less compact shrub and grass canopy, no traces of erosion can be detected.

The shallow terra rossa and brown rendzina soils are well structured, rich in organic matter, and have a high infiltration capacity. Thus, even on fire denuded steep slopes, no signs of runoff and soil movements have been observed after heavy rainstorms in western Galilee. On the other hand, on poorer, more exposed, and on less fertile, highly calcareous slopes disturbed and compacted by cattle and goat grazing, erosion damage after fire has been considerable.[30]

At the same time, both the pale rendzina and terra rossa soils on mountain plateau and moderate slopes which were terraced and cultivated in the distant past, but their terraces neglected, have suffered from severe erosion. After the terrace walls were destroyed and the soil washed away, only a shallow soil layer of 20 cm has been left, exposing the rock surface. Moreover, stone heaps were left in the center of the terraces which, after being exposed by erosion, were no longer carried away to the terrace walls as in ancient times. The vegetation cover is scarce and consists chiefly of *Calycotome villosa* and stunted *P. lentiscus* bushes near the remnants of the terrace walls and of *S. spinosum* and other low valuable woody and herbaceous plants.

V. THE MULTIPLE BENEFITS DERIVED FROM MEDITERRANEAN UPLANDS

At present, the most important benefits derived from the Mediterranean uplands are the conversion of their primary productivity into animal products by uncontrolled grazing and browsing, and into firewood by cutting. At the same time this conversion is the major cause for their degradation. Due to poor silviculture, animal husbandry, and pasture management, the benefits are very low but the effects are often disastrous. Animal and forest products will now be considered, in turn.

A. Livestock Production
According to Le Houeŕou,[23] the total area of pasture land in the Mediterranean Basin is

estimated at about 1.2 million km², including almost the whole of the uncultivable land, comprising 52% of the total land and substantial parts of the remainder used as grain stubble and fallow. With a livestock population of about 370 million sheep equivalents, this corresponds to a stocking intensity of 2.2 sheep per hectare. This is equivalent to a mean production of 2000 kg of dry matter, which is a rather high stocking rate for largely unimproved and mismanaged rangelands.

According to French,[13] the density of goats relative to the human population in the Mediterranean is the highest in the world. They are best adapted to the rugged terrain and can make best use of lignified woody plants by conversion of sclerophylls and thorny plants into milk, meat, hair, and skin.[11] They have become the major source of income in many Mediterranean countries, especially in Greece, Turkey, North Africa, and the Near East.

The reproduction rate of these goats is very high: 100 goats in 5 years from 1 pair as compared to 32 sheep and 10 cattle. They are the most efficient converters of fiber-rich and lignified woody plants and are able to consume daily 8% of their live weight as compared with 3% or less by cattle and sheep.[13]

In typical brushland in Galilee, free ranging Arab Mamber goats browsed on 30 of 40 woody species. The remaining chiefly aromatic subshrubs were eaten occasionally. On the other hand, cattle consumed only ten species of woody plants, chiefly sclerophylls, mainly in the dry season and after fire.[30]

In similar brush ranges in Greece, under controlled grazing management, goats gained 25 kg live weight and yielded 30 ℓ milk per head and annum in Greece, where cattle or sheep would have obtained very little if any fodder.[25]

Most of these woody plants contain only 400 to 600 kcal net energy per kilogram dry matter, as compared to 600 to 800 kcal for green herbaceous pasture plants. However, because of their capacity for tapping moisture surplus from deep rock layers in the summer, they can provide green fodder in the dry season and in drought years, and their leaves are considerably higher in protein, phosphate, and carotene, especially when regenerating after fire, coppicing, or browsing. They are therefore of greatest value in late fall and early winter when herbaceous pasture production reaches its lowest levels.

This is not only the period of greatest nutritional shortage from these herbaceous plants, but also the period when the bare soil is most susceptible to trampling damage, impairing the infiltration capacity and structure, and thereby increasing erosion hazards. Therefore, from the point of view of soil and water conservation, goats, or any other browsing animal, which can exist at this critical time almost solely from woody plants of these shrub covered ranges, have great advantages over cattle and even more over sheep, which tend to congregate in dense flocks.

Practical ways for the improvement of goat husbandry and shrub pastures — including utilization of fenced brush ranges by improved Turkish Angora goats from Texas — have been discussed in detail by Naveh.[33] However, there is little hope for prevention of further degradation without strictly controlled grazing, which can best be achieved in fenced paddocks and ranges. For such rational management policies, public and communal ownership, allowing unrestricted grazing for everybody, are major constraints. As mentioned above, in the past, diseases, starvation in drought years, and fodder shortage in early winter prevented overgrazing during the spring flush season and seed setting. Currently, the ''tragedy of the commons'' is acting as a positive runaway feedback for accelerated pasture depletion. Adverse effects on livestock, shared by all herdsmen and now partly eliminated by supplemental feeding, detracts only to a small degree from the benefits gained by each individual keeping additional animals on common pastures.

The greatest potential for livestock production is in the above-mentioned open woodlands and derived grasslands on the more fertile rendzina and basaltic soils of the xerothermo-mediterranean regions. On such soils, as shown by Gutman and Seligman,[15] manifold increase in pasture and livestock production has been achieved by the replacement of the traditional, uncontrolled year-round open ranching of low quality livestock by more rational and conservative grazing practices, in combination with better livestock husbandry.

In these conditions, at the Kare Deshe experimental farm near the Sea of Galilee, a comparison of heavy and moderate stocking levels over a period of 10 years with a mixed European and local beef cattle herd revealed that although continuous grazing resulted in higher live weight gains, no important differences in cattle production per unit area were observed. What is most relevant to soil protection, early winter rotational grazing — even under heavy stocking rates — improved pasture composition, favored faster growing grasses over nonpalatable herbs, and left more litter and ungrazed herbage at the beginning of the main rainy season.

At the Neve Yar experimental station in Lower Galilee, rotational deferred grazing over a period of 6 years improved pasture composition and doubled the pasture production. By additional fertilizing and selective weed control, pasture output increased threefold to 3000 kg/ha/year of dry matter and to 200 kg live weight increase per hectare. The great dependence of pasture production on seasonal and annual climatic fluctuations cannot only be overcome, but production can be increased by intensive improvement.[37]

The dramatic rises in fodder production and quality, combined with soil protection and improvement of rocky hillsides, can be achieved not only in experimental stations, but also by typical Arab hill farmers under professional guidance. This has been demonstrated in the Samarian and Judean mountains on rocky and shallow hillsides by reseeding of a drought-resistant strain of Bur Clover, *Medicago polymorpha,* in combination with superphosphate dressing and deferred grazing.[4] Similar results were also achieved in Cyprus on marginal, arable land, with milk sheep, where grass pasture improvement by reseeding of annual legumes and phosphating was combined with improvement of carob fodder trees.[21]

We may summarize by stating that the most critical period for soil and water erosion is in late fall and early winter, when herbaceous pasture production is at its lowest and susceptibility of the soil as well as of the newly emerging pasture plants are at their highest. These damages are greatest in drought spells and in drought years but, as we have seen, the great seasonal and annual fluctuation in herbaceous pasture production cannot be eliminated, even by intensive range improvement. On the other hand, this major bottleneck in livestock production and pasture stability and productivity could be overcome by the enrichment of herbaceous pastures, and the replacement of low valuable woody plants by more palatable and productive, summergreen, and drought-tolerant shrubs and trees, which would provide nutritious fodder in fall and winter and in drought years. Plants such as the indigenous drought- and fire-resistant, root regeneration oak and pistacia species, are efficient in tapping of energy, water, and nutrients from the shallow soil and underlaying rock layers, and therefore valuable in soil and water conservation. However, they should have much higher levels of efficiency in the channeling of solar energy into economically useful animal and other products.

Among the indigenous plants, the carob tree is of highest value in this respect, but several exotic trees and shrubs, studied in long-term introduction trials in northern Israel, show great promise for this purpose. Most outstanding, because of their adaptability to poor rocky and calcareous soils, and their great palatibility and regeneration potentials after heavy browsing in October through December, have been *Atriplex nummularia,* one of the most valuable native fodder shrubs in Australia, and *Cotoneaster franchettii,* which is also a most valuable ornamental plant.

The establishment of these and other promising fodder shrubs and trees on depleted scrubland, covering many thousands of hectares of Mediterranean uplands, could be achieved by the same relatively inexpensive techniques used for pine afforestation on these uplands. This could lead to a manifold rise of their carrying capacity and productivity, and would have dramatic results, especially if combined with the improvement of livestock husbandry, and if carried out as part of a master plan for multiple upland reclamation and afforestation.

B. Forestry Production

Le Houerou[23] has summarized the effects of erosion, based on comparison of aerial photographs taken at intervals of 10 to 30 years. He reached the conclusion that the amount of erosion is inversely proportional to the degree of afforestation — higher in the south and east of the Mediterranean than in the north and west. There are, however, exceptions, such as in the badly eroded southern Apenines. In general 60 to 70% of the annual erosion occurs between September and November. The amount of erosion in comparable areas is 50 times higher on bare soil than under well-developed forest cover and the amount of water runoff is five times higher.

In North Africa, the amount of erosion ranges from 2 to 3 t/ha/year in small wooded catchments to 20 t/ha/year in cleared catchments on friable rock (marl hills), and on larger catchments with little forest, from 5 to 10 t/ha/year. In certain cases of large cleared catchments with extensive exposures of marl, these may reach even terrible maxima of 30 or even 53 t/ha/year in Algiers, corresponding to erosion of a layer of soil 0.2- to 0.4-mm thick over the catchment. In Algiers as a whole, Greco[14] estimates that 40,000 ha of arable land, or 0.06% of the existing land area, are lost every year.

Out of 207 million acres of forest land in the Mediterranean Basin only 50 million are reasonably stocked with trees. On 100 million acres only traces of forest vegetation remain of the original cover. The importance of these forests in general, and especially for fuel for the local population, has been described in detail by Thirgood.[51] In recent years, with the sharp increase in the cost of fossil energy as fuel, the consumption of firewood has increased considerably and, according to Le Houerou,[23] 50 million people in the Mediterranean are cutting the equivalent of 27 million ha/year of forests.

Coniferous trees, especially pines, play an important role in afforestation projects because of their hardiness, easy establishment, and rapid growth. In countries such as Israel, Cyprus, Spain, and Greece, such upland pine afforestations have saved thousands of hectares from further depletion, provided labor for rural populations, and dramatically improved the denuded and barren mountain landscapes. In many cases they opened the way for the vegetative regeneration of the sclerophyll understory. This could be utilized by proper forest management for the creation of seminatural, mixed, and multilayered multipurpose forest. However, as has been rightly pointed out by Mooney and Gulmon,[26] primary production in a habitat, and therefore also wood production, is determined by the resources available at the site, independent of the species utilizing it.

Pinus halepensis, as well as *P. brutia* are well adapted to the poor water and nutrient resources available on the drier and rocky slopes, and are used chiefly for afforestation in the Mediterranean; this is due to their development of an extensive root system. According to El Aouni,[10] the young trees use 27%, and the adult trees use 45% of the total assimilates. This is in addition to the high carbon synthesis cost of secondary lipid metabolites, and the production of the numerous cones and large amounts of seeds, which are apparently necessary to ensure successful natural establishment after fire.

For these reasons the annual wood increment per hectare from pine forests on these uplands is less than 1.00 m³, and on the rockier and drier slopes even less than 0.5 m³. However, even these low economic benefits, as well as their social values for recreational uses, are impaired by the great fire hazards and the heavy expense of their prevention and of fighting wildfires.

The early rapid development of such hardy pioneer trees has created the mistaken impression that foresters can fool nature in the Mediterranean and establish dense, stable, and productive pine forests such as those in much more favorable conditions in central Europe.

However, even their present low productivity is now severely threatened by combined and synergistic impacts of photochemical air pollutants and pest infestations. Recent field and laboratory studies[41] revealed that widespread chlorotic mottle and decline is being caused in Israel, as in California, by atmospheric ozone concentrations above 0.05 ppm, followed in *P. halepensis* by severe infestations of *Matsucoccus josephi* scales. Most recently, ozone-stressed *P. canariensis* trees are also being attacked and killed by bark beetles in the coastal region, where ozone levels already exceed 0.1 ppm. Similar damages can be expected in all other rapidly developing Mediterranean countries. They have been observed in Greece, in a forest above Thessaloniki, planted to *P. brutia,* where the sea breeze carries photochemical oxidants from urbanized, motorized, coastal regions into forested slopes. This will doubtless increase everywhere from year to year.

In younger pine stands there is still much grazable forage, but with increasing shade and accumulation of very slowly decomposing needle litter, palatable pasture plants are diminishing, together with edible plants, flowering geophytes, and wildlife. Because grazing is mostly uncontrolled, it is regarded by foresters as a menace and not an asset. On the other hand, local populations who have been deprived of their grazing and hunting grounds without gaining any direct benefits from the new forests, are largely hostile and noncooperative. This may be expressed in wanton destruction of forests by fire with disastrous results.

Morandini[27] points to the renewed importance of sclerophyll forests and shrublands for local firewood supply and, by mechanized wood chopping, for commercial fiber production. This could pay for the thinning and opening up of dense shrubland thickets into woodlands and recreation forests. He opposes the conservation policy of a total halt of intervention out of entirely theoretical considerations of a hypothetical forest "climax". He feels that the natural place for intensive tree cultivation in this zone is outside the forest. Plantations of high yielding timber and fiber trees, such as poplar and eucalyptus, should be established in the fertile, deeper, lowland soils, or on the gentle slopes of abandoned agricultural land suitable for more economical forest production with mechanized cultivation.

Additional forestry production sources of direct economic benefit, such as cork from *Quercus suber,* and from the collection of aromatic plants as spices, balsam, and medicinal teas, are steadily decreasing with the advance of urbanization and the resulting depopulation, chiefly in the western Mediterranean uplands. This is in spite of the fact that there is a growing demand for many natural health products.

VI. OTHER BENEFITS AND FREE ECOLOGICAL SERVICES

Without doubt, the greatest indirect social and economical benefit derived from these uplands stem from their amenity and scenic values, which are enjoyed not only by the local populations, but by many millions of tourists. This is true not only for the seashores but also for these uplands, with their extraordinary wealth of plant and animal life, especially in the spring, when hundreds of colorful flowering plants can be found and, unfortunately, are also picked and even sold, and their bulbs exported.

Of no lesser importance are the "free ecological services" these uplands provide as stable, seminatural life-supporting systems, as buffer zones, and "living sponges" for protection of watersheds, while aiding in the control of floods, erosion, and environmental pollution in the densely populated coastal regions and intensively cultivated areas. In similar conditions in California, Westman[56] has shown that these functions are worth many millions of dollars if modern technology should have to replace them or repair the damage of their destruction.

Of special importance are the open xerothermic and semiarid woodlands, shrublands, and grasslands as the last refuge for a wide variety of plant and animal species. Studies in Israel[40] have revealed that their great floral diversity include not only highly attractive, ornamental geophytes, but also progenitors of many domesticated cereals and pulses, and a wealth of hardy and drought-resistant herbaceous and woody plants which could serve as genetic stock for future evolution and economic uses for food, fodder, industrial, pharmaceutical, cosmetic, ornamental, as well as vegetation engineering, slope protection, and other purposes. Their loss, not only by further uncontrolled depletion and overuses, but also by complete protection in order to restore an illusionary climax, or by planting dense pine or eucalyptus forests with highly doubtful economic value in these conditions, would be final and irrevocable.

The continuation of these vital ecological and evolutionary functions cannot be ensured, either in botanical gardens or in small completely protected nature reserves. The latter soon become very dense, unpenetrable, monotonous, and species poor, but highly fireprone brush thickets with low biological, scenic, and recreational values. These values can best be ensured by imposing strictly controlled optimum defoliation pressures by grazing and browsing of livestock and/or wildlife, especially ungulates, as well as by periodic coppicing and, if necessary, also by controlled burning. In this way further economic benefits could be derived, while ensuring at the same time the functional integrity of these ecosystems and their "noneconomic richness".

We may therefore conclude that there is urgent need for integrative strategies of multiple benefit upland utilization in which all these biological, ecological, sociological, and economical functions of the landscape could be optimized for maximum overall benefits, including soil and water conservation.

In order to ensure the greatest possible overall benefit from these often-clashing land use demands, flexible multipurpose strategies should be applied according to site potentials and local and national requirements. This could be realized in practice by the creation of closely interwoven networks of multiple land use patterns. For this purpose, flow charts of management strategies and cybernetic models of the mutual impacts of these options and their production, protection, and regulation functions and other environmental variables should be prepared.[35,36]

In nature reserves the main aim should be the conservation of biological productivity, diversity, and stability through scientific dynamic conservation management. In nature parks and recreation areas it should be the optimization of landscape, wildlife, and recreation amenities to enable maximum enjoyment with minimum damage to natural and cultural resources; in the remaining open uplands, used primarily for socioeconomic benefit, two major "ecotechniques" should be applied.

1. Vegetation and ecosystem management of the existing plant cover and controlled manipulation of the soil-plant-animal complex by controlled grazing, thinning, coppicing, and pruning of trees, and, if necessary, controlled fire and pastoral agrotechnical improvements, such as reseeding, fertilizing, selective weed control, according to site potentials and socioeconomic and other requirements and conditions.

2. Multipurpose afforestation and reforestation. This is essential, wherever land denudation has reached an advanced stage, as well as in man-created habitats of roadsides, camping grounds, and recreation sites. It should also include the establishment of fire protection strips and buffer zones for urban and industrial complexes. In contrast to the single-layered, monotonous, and vulnerable conifer plantations, the aim is the creation of a new type of seminatural, multilayered, and stable park-forest with multiple ecological, economical, and social benefits. These new forests should resemble in structure, diversity, and stability the richer multilayered, park-like, and semiopen Mediterranean forests and woodlands with a scattered tree overstory, like the previously described thinned and pruned oak and mixed sclerophyll recreation forest in northern Israel. At the same time, however, they should be of greater economical and ornamental value by the replacement of inferior indigenous woody plants by more valuable introduced shrubs and trees.

Although it is still too early for any final conclusions, there are already available a number of highly promising plants for these purposes with multiple use values which can be established and maintained like pine trees with minimum care, and without any supplemental irrigation. Some of these are used already in afforestation projects in Israel.

This integrative approach is also essential for the urgently needed interdisciplinary ecosystem research on the ecological and economical implications of different kinds and intensities of these land uses, and especially of recreational use and their combinations, to enable the quantification of aforementioned models. Foresters should play an important role in applying their rich silvicultural experience to these new approaches in the introduction and establishment of suitable, low flammable fodder, forest trees, and shrubs, and in the development of dynamic fuel management, including prescribed burning.

However, the introduction of sounder, more rational, and conservative management and improvement practices described earlier, are facing severe socioeconomic constraints. The first and foremost is the public and communal ownership or tenure of the greatest parts of all Mediterranean uplands, allowing unrestricted grazing and cutting for everybody.

Another constraint is the general neglect and indifference of politicians, decision makers, their advisors, and economists, toward these untillable Mediterranean uplands. In general, all development and improvement schemes are directed to the more productive agricultural lands, where the expected short-term benefits appear to be much more favorable. If there are any efforts and investments made on these uplands, they are directed toward timber production and commercial recreational uses, with little direct benefit for local and chiefly pastoral users.

As described in detail in Naveh and Licberman,[42] grazing control and the rise in upland production in general can be achieved only as part of a new cultural feedback by comprehensive overall improvement and development of the Mediterranean total human ecosystem. This includes the rise in the socioeconomic status of the pastoralist, coupled with the elimination of ignorance and indifference at the local, communal, regional, and national level. It requires, above all, a radical change in the attitude of decision makers, land planners, owners, and users from shortsighted exploitation to more farsighted landscape ecological determinism, leading to comprehensive and long-term master plans. In these, the needs for conserving and improving these uplands and their soil, water, and vegetation resources should be reconciled with socioeconomic advancement and dynamic conservation, as described earlier. These should serve not only the local livestock industry, or the national forest production and other sectorial economic interests, but should constitute an integral part of multiple benefit upland development and rehabilitation. It should be realized that these changes in attitudes and concepts cannot be initiated *merely* by scientific and technological means, but *only* by a coordinated effort for public environmental education at all levels.

In conclusion, there is urgent need for an integrative approach to reconcile the need for conserving the scarce soil and water resources as well as the biological diversity and productivity of these upland ecosystems, with the socioeconomic needs for its inhabitants, and the national economy. This requires the cooperation of all those who care for, those who deal with, and those who live from its resources. Therefore, the chief actions required are in the realms of interdisciplinary system-oriented landscape-ecological education, planning, management, development, afforestation and research aimed at the conservation and restoration of the seminatural, nontillable, Mediterranean landscapes.

REFERENCES

1. The Mediterranean (special issue), *Ambio,* 6, 1977.
2. **Banlieu, J. L.,** Analyse pollinique dans les Mont hispanica, *Inst. Fr. Invest. Exp.,* 20, 24, 1969.
3. **Bradbury, D. E.,** The physical geography of the Mediterranean lands, in *Ecosystems of the World II, Mediterranean-Type Shrublands,* diCastri, F., Goodall, D. W., and Specht, R. L., Eds., Elsevier, Amsterdam, 1981, 53.
4. **Briegeeth, A. M.,** Pasture and range areas in Judea and Samaria, in Proc. Int. Symp. on Pastoral Sheep Farming Systems in Intensive Economic Environments, Jerusalem, Ministry of Agriculture, Tel Aviv, 1981.
5. **Catarino, F. M.,** International Symposium on *Ceratonia siliqua* L., Centro de Engenharia Biologia das Universidades, Lisbon, 1981.
6. **Dan, J.,** The disintegration of Nari Lime crust in relation to relief, soil and vegetation, *ITC,* 189, 1963.
7. **Dan, J. and Yaalon, D. H.,** Trends of soil development with time in the Mediterranean environments of Israel, in Trans. Int. Conf. Medit. Soils, Madrid, 1966, 139.
8. **Danziger, Y., Morin, J., and Naveh, Z.,** *The Rehabilitation of the Nesher Quarry,* Technion, Haifa, Israel, 1972.
9. **diCastri, F.,** Mediterranean-type shrublands of the world, in *Ecosystems of the World II, Mediterranean-Type Shrublands,* diCastri, F., Goodall, D. W., and Specht, R. L., Eds., Elsevier, Amsterdam, 1981.
10. **El Aouni, M. H.,** Percessus Determinant la production du Pin d'Alep *(Pinus halepensis)* Mill.: Photosynthese, Croissance et Repartition des Assimilates, Ph.D. thesis, University of Paris, Paris, 1980.
11. Food and Agriculture Organization, Report on Goat-Raising Policies in the Mediterranean and Near East Regions, Experimental Program of Technical Assistance, FAO Publ. No. 1929, FAO, Rome, 1965.
12. **Fosberg, F. R.,** Restoration of lost and degraded habitats, in *Future Environments of North America,* Darling, F. F. and Milton, J. P., Eds., Natural History Press, New York, 1956, 503.
13. **French, M. H.,** Observations on the Goat, FAO Agric. Stud. No. 80, Food and Agriculture Organization, Rome, 1970.
14. **Greco, J.,** L'Erosion, l'Defense et la Restauration des Sols en Algerie, Ministry of Agriculture, Algiers, 1966.
15. **Gutman, M. and Seligman, N.,** Grazing management of Mediterranean foothill range in the Upper Jordan River Valley, *J. Range Manag.,* 32, 86, 1979.
16. **Heichelheim, F.,** Effects of classical antiquity on the land, in *Man's Role in Changing the Face of the Earth,* Thomas, W. L., Ed., University of Chicago Press, Chicago, 1956, 165.
17. **Hallam, S. J.,** *Fire and Hearth. A Study of Aboriginal Usage and European Usurpation in South-Western Australia,* Australia Institute of Aboriginal Studies, Canberra, Australia, 1979.
18. **Horowitz, A.,** *The Quarternary of Israel,* Academic Press, New York, 1979.
19. **Iversen, J.,** Forest clearance in the Stone Age, in *Man and the Ecosphere,* Ehrlich, E. R., Holdren, J. P., and Holm, R. W., Eds., W.H. Freeman, San Francisco, 1971, 26.
20. **Jenny, H.,** *The Soil Resource, Origin, and Behavior,* Springer-Verlag, Berlin, 1980.
21. **Jones, D. K.,** Combining pasture improvement and carob production in Cyprus, *J. Range Manag.,* 8, 151, 1955.
22. **Katznelson, J.,** The variability of rainfall in Palestine and the statistical methods of its measurements, *Isr. Meteorol. Serv. Ser. E,* 4, 32, 1956 (in Hebrew).
23. **Le Houerou, H. N.,** Impact of man and his animals on Mediterranean vegetation, in *Ecosystems of the World II, Mediterranean-Type Shrublands,* diCastri, F., Goodall, D. W., and Specht, R. L., Eds., Elsevier, Amsterdam, 1981, 479.
24. **Lewis, H. T.,** *Pattern of Indian Burning in California: Ecology and Ethnohistory,* Ballena Press Anthropol. Pap. No. 1, Ballena Press, Ramona, Calif., 1973.
25. **Liacos, L. G. and Mouloupoulos, C.,** *Contribution to the Identification of Some Range Types of Quercus coccifera,* University of Thessaloniki, Thessaloniki, Greece, 1967, 54.
26. **Mooney, H. A. and Gulmon, S. L.,** The determinants of plant productivity — natural versus man-modified communities, in *Disturbance and Ecosystems,* Mooney, H. A. and Godron, M., Eds., Springer-Verlag, Berlin, 1983, 146.
27. **Morandini, R.,** Problems of Conservation, Management, and Regeneration of Mediterranean Forests: Research Priorities, MAB Tech. Notes 2, United Nations Educational, Scientific, and Cultural Organization, Paris, 1977.
28. **Morin, J.,** Mechanized biological stabilization of slopes, in Symp. on Vegetation Engineering, Tel-Aviv, 1968, 19.
29. **Narr, K. J.,** Early food-producing populations, in *Man's Role in Changing the Face of the Earth,* Thomas, W. L., Ed., University of Chicago Press, Chicago, 1956, 134.
30. **Naveh, Z.,** Agro-Ecological Aspects of Brush Range Improvement in the Maquibelt of Israel, Ph.D. thesis, Hebrew University, Jerusalem, 1965 (in Hebrew with English summary).

31. **Naveh, Z.,** Mediterranean ecosystems and vegetation types in California and Israel, *Ecology,* 48, 445, 1967.
32. **Naveh, Z.,** The ecology of fire, Proc. Annu. Tall Timbers Fire Ecol. Conf., Tallahassee, Fla., March 1973, 131.
33. **Naveh, Z.,** The ecological management of non-arable Mediterranean upland, *J. Environ. Manag.,* 2, 351, 1974.
34. **Naveh, Z.,** Degradation and rehabilitation of Mediterranean landscapes, *Landscape Planning,* 2, 133, 1975.
35. **Naveh, Z.,** A model of multi-purpose ecosystem management for degraded Mediterranean uplands, *J. Environ. Manag.,* 2, 31, 1978.
36. **Naveh, Z.,** A model of multi-purpose management strategies of marginal and untillable Mediterranean upland ecosystems, in *Environmental Biomonitoring, Assessment, Prediction and Management — Certain Case Studies and Related Quantitative Issues,* Cairns, J., Patil, G. P., and Waters, W. E., Eds., International Cooperative, Fairland, Md., 1979, 269.
37. **Naveh, Z.,** The dependence of the productivity of a Mediterranean hill pasture ecosystem on climatical fluctuations, *Agric. Environ.,* 7, 47, 1982.
38. **Naveh, Z.,** The vegetation of the Carmel and Nahal Sefunim and the evolution of the culture landscape, in *The Sefunim Cave on Mt. Carmel and Its Archeological Findings,* Ronen, A., Ed., B.A.R. Int. Ser., Oxford, 1984.
39. **Naveh, Z. and Dan, J.,** The human degradation of Mediterranean landscapes in Israel, in *Mediterranean-Type Ecosystems, Origin and Structure, Ecological Studies: Analysis and Synthesis,* Vol. 7, diCastri, F. and Mooney, H. A., Eds., Springer-Verlag, Berlin, 1973, 373.
40. **Naveh, Z. and Whittaker, R. H.,** Structural and floristic diversity of shrublands and woodlands in northern Israel and other Mediterranean areas, *Vegetation,* 41, 171, 1979.
41. **Naveh, Z., Steinberger, E. H., Chaim, S., and Rotmann, A.,** Photochemical air pollutants — a threat to Mediterranean coniferous forests and upland ecosystems, *Environ. Conserv.,* 2, 301, 1981.
42. **Naveh, Z. and Lieberman, A. S.,** *Landscape Ecology — Theory and Applications,* Springer-Verlag, Berlin, 1984, 256.
43. **Papanastasis, V. P.,** Effects of season and frequency of burning on a phryganic rangeland in Greece, *J. Range Manag.,* 33, 252, 1980.
44. **Papanastasis, V. P. and Liacos, G.,** Productivity of hermes oak brushlands for goats, *Int. Symp. on Browse in Africa,* International Centre for Africa, Addis Ababa, Ethiopia, 1980.
45. **Perles, C.,** *Prehistore du Feu,* Masson, Paris, 1977.
46. **Polianos, A. N.,** *Die Hoeshle der Petralonischen Archanthropinen,* Anthropological Society of Greece, Athens, 1982, 706.
47. **Reifenberg, A.,** The struggle between the desert and the sown, in *Rise and Fall of the Levant,* Jewish Agency, Jerusalem, 1951.
48. **Ron, Z.,** Agricultural terraces in the Judean mountains, Israel, *Isr. Exp. J.,* 16, 33, 1966.
49. **Schiechtl, H.,** *Bioengineering for Land Reclamation and Conservation,* University of Alberta Press, Edmonton, Alberta, Canada, 1980.
50. **Taylor, F. H.,** The destruction of the soil in Palestine, *Bull. Soil Conserv. Board, Palestine,* 2, 1946.
51. **Thirgood, J. V.,** *Man and the Mediterranean Forest. A History of Resources Depletion,* Academic Press, London, 1981.
52. **Tomaselli, R.,** Degradation of the Mediterranean maquis, in *Mediterranean Forests and Maquis: Ecology, Conservation and Management,* MAB Tech. Notes 2, United Nations Educational, Scientific, and Cultural Organization, Paris, 1977, 33.
53. **Trabaut, L.,** Man and fire: impacts on Mediterranean vegetation, in *Ecosystems of the World II, Mediterranean-Type Shrublands,* diCastri, F., Goodall, D. W., and Specht, R. L., Eds., Elsevier, Amsterdam, 1981, 523.
54. United Nations Educational, Scientific, and Cultural Organization, Bioclimatological Map of the Mediterranean Zone. Arid Zone Research, UNESCO, Paris, 1963.
55. **Vita-Finzi, C.,** *The Mediterranean Valleys; Geological Changes in Historical Times,* Cambridge University Press, Cambridge, England, 1969.
56. **Westman, W. E.,** How much are nature's services worth?, *Science,* 197, 960, 1977.

Chapter 6

ENGINEERING MEASURES: WATERWAYS AND DIVERSION CHANNELS

Herman J. Finkel

I. INTRODUCTION

There is a certain logical order in applying soil and water conservation measures to cultivated fields, proceeding from the simple and less expensive to the more complex and expensive. Only where the lesser measures are inadequate are the greater measures added. Note the word "added". The engineering measures are not a substitute for the agronomic measures previously described, but should be superimposed upon them. A full program begins with correct land use (insofar as economic realities permit), proceeds to selection of allowable intensity of rotation, and is followed by contour farming, with strip cropping where needed. All of these measures are within the farmer's ability to establish alone, with the advice of a competent farm planner and a bit of technical help. If all of these practices are not sufficient, the farmer must turn to the engineering measures which require a somewhat higher level of technical assistance, and are more costly. These measures are, in themselves, graduated from the simpler to the more complex, and should be applied in that order, which is more or less the sequence of the presentation in this chapter.

II. PROTECTED WATERWAYS

The first engineering measure to be established is the protected, or permanent, waterway. This is, in effect, both an agronomic and an engineering measure. A field on rolling topography can naturally be divided into ridges and depressions (see Figure 1). The depressions, or draws, form natural outlets or waterways for the disposal of runoff. If they are spaced too closely, not all of them need be used, and only some may be selected for conversion into waterways. A reasonable spacing of the waterways is from about 75 to 200 m, depending upon the slope of the land and the general unevenness of the topography. Too close a spacing will result in very short contour rows, and will retire too much land area from cultivation. With a wider spacing, the smaller rills and draws between the selected waterways can be filled in as much as possible, if they are not too deep, and included in the contour-cultivated portions of the field.

Once the site of a waterway has been selected, a channel of the proper dimensions must be shaped in the draw. This must satisfy two requirements: it must be adequate to pass the discharge of the design flood, and the velocity of flow must be low enough to prevent channel scour or erosion. These factors will be considered in turn.

A. The Design Flood

The maximum rate of runoff from a small watershed is not easy to determine. The inadequacies of the rational runoff formula were discussed in Chapter 2. In addition, there is, especially in Third World countries, a dearth of long-term rainfall intensity data. It is simpler, therefore, to use one of the empirical formulas described in Chapter 2. For example, in the dry, Negev region of Israel, the following formula was developed for average peak flow, $\overline{Q}$:

$$\overline{Q} = 0.5 \, A^{0.67} \tag{28}$$

where A is the drainage area in square kilometers and $\overline{Q}$ is in cubic meters per second.

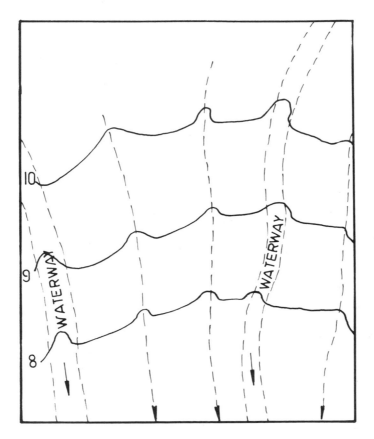

FIGURE 1.   Selecting certain draws for waterways.

The question which immediately arises is for what probability of recurrence should a waterway be designed? This question, in fact, arises for every type of hydraulic structure that carries flood flows. The problem is one of economics. The cost of designing for a maximum flood of infrequent occurrence must be balanced against the damage which might be caused by the flood exceeding the capacity of the structure on the average of once in a number of years. This damage may be to the structure itself, or to the adjoining property. In the case of a grassed waterway in the natural depression of an agricultural field, inadequate design for discharge would cause temporary overtopping, but very little permanent damage to the channel itself, or to adjacent fields which could not be repaired at a relatively low cost. Consequently, there is little justification in selecting a design flood which has an average recurrence frequency of over 5 years, or a 20% probability of occurrence within a given year. Some designers would go as low as the so-called 2-year flood, or 50% probability.

Referring again to the Israel example, the 50% probability for peak discharge is given by the expression

$$Q = 0.24 \; A^{0.67} \tag{29}$$

Thus, if the waterways are spaced 200 m apart and the length of the waterway is, for example, 1000 m, the drainage area would be 0.2 km^2, and the design flood would be 0.08 m^3/sec (2.8 ft^3/sec). The long-term average annual peak flow would be

$$Q = 0.5 \; A^{0.67} = 0.16 \; \text{m}^3/\text{sec} \; (5.6 \; \text{ft}^3/\text{sec}) \tag{30}$$

Table 1

MEAN VELOCITY VALUES AT INCIPIENT SCOUR

| Type of soil | Permissible mean velocity V | | | |
| | Clean water | | Water containing colloids | |
	m/sec	ft/sec	m/sec	ft/sec
Very fine sand	0.45	1.48	0.75	2.46
Sandy loam	0.55	1.80	0.75	2.46
Silty loam	0.60	1.97	0.90	2.95
Alluvial silt, without colloids	0.60	1.97	1.00	3.28
Dense clay	0.75	2.46	1.00	3.28
Hard clay, colloidal	1.10	3.61	1.50	4.92
Very hard clay	1.80	5.90	1.80	5.90
Fine gravel	0.75	2.46	1.50	4.92
Medium and course gravel	1.20	3.94	1.80	5.90
Stones	1.50	4.92	1.80	5.90

This has a frequency of about once in 3 years, which should be sufficient protection for the typical grassed waterway. (It should be noted that in this example, the watershed area was calculated only from the cultivated field which drains into the waterway along the contour lines or through a system of terraces. If there is additional drainage area from above this must be taken into consideration.)

B. Allowable Velocity

The next factor to be considered is the allowable velocity of flow. The safe velocity on bare earth channels varies with the soil type, and various recommendations are presented in the technical literature. One example is given in Table 1.

If the channel is protected by a permanent grass cover, the allowable velocities may be higher. The effect of the grass is twofold: on the one hand it permits a higher safe velocity; on the other, it reduces the discharge capacity of the channel by reducing the cross section and by increasing friction. The hydraulic characteristics of grass cover will be treated in more detail in the following section.

C. Channel Design

The hydraulic design of waterway channels is usually done with the well-known Manning formula, which is

$$V = (1/n)(R^{0.67} S^{0.5}) \tag{31}$$

where V = velocity of flow, in meters per second, R = A/P, hydraulic radius, in meters, A = cross sectional area in square meters, P = wetted perimeter, in meters, and S = channel gradient, expressed as a decimal.

In the FPS system the formula is

$$V = (1.486/n)(R^{0.67} S^{0.5}) \tag{32}$$

where V = feet per second, and R = feet. In both systems, n is the Manning roughness coefficient which, for bare earth, is about 0.03, and for grassed surfaces, is variable from 0.07 to 0.10. Many standard texts on open channel hydraulics give tables of numbers to the exponents 0.67 and 0.5. With the advent of pocket calculators such tables are no longer necessary.

The corollary formula which must be employed to complete the design is

$$Q = AV \tag{33}$$

where Q is the discharge in cubic meters or cubic feet per second, A is the cross sectional area of flow in square meters or square feet, and V is the average velocity in meters or feet per second.*

In the special case of grassed waterways, channel capacity is usually not the limiting factor. The main problem is to keep the velocity within the allowable to prevent channel scouring. This may be accomplished with a wide, shallow channel having a minimal R, as may be seen in the following examples.

1.   What are the capacity and velocity of flow in a channel on a 5% gradient with the cross section shown in Figure 2?

$$A = 3.95 \times 0.15 = 0.6 \text{ m}^2$$

$$P = 4.45 \text{ m}$$

$$R = 0.6/4.45 = 0.135 \text{ m}^2 \text{ R}^{0.67} = 0.26$$

$$S = 0.05$$

$$S^{0.5} = 0.223$$

For bare soil assume n = 0.04

$$V = 1/0.04 \ (0.26 \times 0.223) = 1.46 \text{ m/sec}$$

$$Q = AV = 0.6 \times 1.46 = 0.88 \text{ m}^3/\text{sec} \tag{34}$$

This channel would not be satisfactory because the velocity is too high for bare soil.

2.   Is the channel shown in Figure 3, on a 3% slope, satisfactory for the discharge of 0.16 m³/sec? Assume n = 0.04 for bare earth.

Assume n = 0.04 for bare earth

$$A = 8.12 \times 0.04 = 0.325 \text{ m}^2$$

$$P = 8.24$$

$$R = A/P = 0.04$$

$$V = 1/0.04 \ (0.04^{0.67} \times 0.03^{0.5}) = 0.50 \text{ m/sec}$$

$$Q = AV = 0.325 \times 0.5 = 0.16 \text{ m}^3/\text{sec} \tag{35}$$

This channel is satisfactory both for discharge capacity and average velocity.

It will be noted that for relatively wide and very shallow channels the hydraulic radius, R, is almost equal to the average depth of flow. With the approximation that P, the wetted perimeter of the cross section, is equal to the total width, W, then R is A/W. It is now simple to arrive at the correct dimensions for the channel section. If the design discharge

---

*   The example taken from Ree and Palmer[4] of *Cynadon dactylon* is a very tough and aggressive grass, which, although excellent for erosion control in waterways, is difficult to eradicate in a cultivated field.

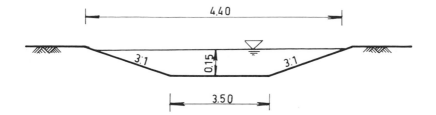

FIGURE 2. Channel section for the first example.

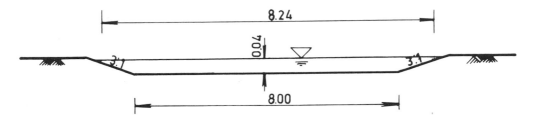

FIGURE 3. Channel section for the second example.

is 0.16 m³/sec, and the allowable velocity is, for example, 0.5 m/sec, the roughness factor, n = 0.04, for bare earth, and the slope of the bed is 3%, we get the following:

$$0.5 = \frac{1}{0.04} R^{0.67} 0.03^{0.5}$$

$$R = \left(\frac{0.5 \times 0.04}{0.173}\right)^{1.5} = 0.04 \qquad (36)$$

The channel should be 0.04-m deep. Now, A = Q/V = 0.16/0.5 = 0.32. The top width of the channel will be W = A/R = 0.32/0.04 = 8 m. With an electronic pocket calculator which gives y to any exponent x, these calculations can be made in a matter of seconds without the need for tables.

The wide, flat trapezoidal section is typical for grassed waterways. However, in actual practice, even though such a section is constructed, it will soon become transformed by the flowing water into something approaching a parabolic section. It is possible, therefore, to design the channel from the very beginning as a parabola. The properties of a wide, flat parabola are shown in Figure 4.

A design for the same conditions as in the previous example is as follows:

$$R = 0.04$$

$$h = 1.5 \times 0.04 = 0.06$$

$$W = 1.5 \, A/h = 1.5 \times 0.32/0.06 = 8.00 \text{ m} \qquad (37)$$

In other words, the same top width as in the flat trapezoidal section will be used on the parabolic section.

In the above examples, it will be noted that for a bare earth channel, a wide, shallow section is needed to keep the velocity of flow within the permissible. As the gradient increases, the required width also becomes larger. This represents a considerable loss of productive area in the field, as well as a high first cost for constructing the section to the required degree of accuracy. The addition of grass cover on the channel surface will have

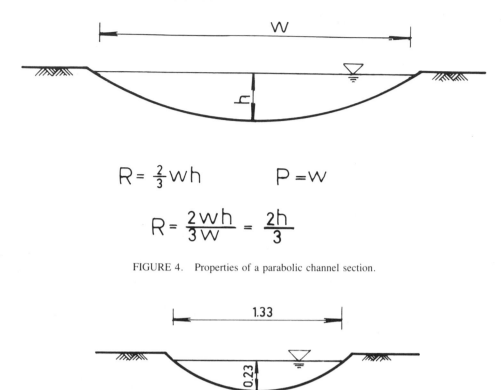

$$R = \tfrac{2}{3} wh \qquad\qquad P = w$$

$$R = \frac{2wh}{3w} = \frac{2h}{3}$$

FIGURE 4.   Properties of a parabolic channel section.

FIGURE 5.   Channel section for the third example.

two effects: on the one hand, it will increase the roughness factor, n (thereby reducing both the velocity and the hydraulic capacity of the cross section), and on the other, it will increase the allowable nonerosive velocity. As an illustration, let us assume that a soft, low-growing grass would have the coefficient n = 0.06, and that the allowable velocity would be 0.80 m/s. Assuming all other conditions are as in the previous example, the calculation of the section would be as follows:

$$Q = 0.16 \text{ m}^3/\text{sec} \quad V = 0.8 \text{ m/sec} \quad S = 0.03$$

$$A = 0.16/0.8 = 0.20 \text{ m}^2$$

$$R = \left\{ \frac{0.8 \times 0.06}{0.173} \right\}^{1.5} = 0.15 \text{ m}$$

$$W = A/R = 0.20/0.15 = 1.33 = 1.33 \text{ m}$$

$$h = 1.5 \quad R = 0.225 \text{ m} \tag{38}$$

The correct parabolic section would be 0.22-m deep at the center and 1.33-m wide at the top (see Figure 5). The effect of the grass cover is thus seen to be very great.

The question now arises as to the actual values of n for different types of grass covers, as determined experimentally. This has been thoroughly investigated by Ree and Palmer.[4] In a series of full-scale hydraulic tests on various types of channel sections, slopes, and grass covers, it was found that the roughness factor, n, is not a constant for a given species of grass but varies with the depth and velocity of the flow. An important factor is the degree of submergence of the grass. When the depth of flow is less than the normal height of the

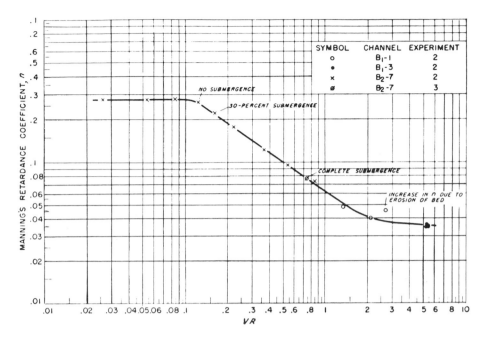

FIGURE 6. Relation of Manning's n and VR for channels lined with long-dormant Bermuda grass.

grass, the water seeps between the close-growing stems and the resistance to flow is very high. As the depth increases more of the grass is submerged, and for many species the grass tends to bend over in the direction of flow, giving a lower n. After complete submergence, the water flows smoothly over the flattened grass and the n value tends to approach a constant which is quite low. The empirical data was found to correspond quite closely to a set of curves of n as a function of VR, where these variables are as previously defined.

A typical n-VR curve for channels lined with Bermuda grass (*Cynadon dactylon*) is shown in Figure 6. Many other n-VR curves are presented by Ree and Palmer[4] for different grasses in various stages of growth, and data are included for grasses which may have similar characteristics to those grown in particular regions of concern.*

D. Establishing the Grassed Waterway

To establish a grassed waterway, two operations are required: shaping and grassing. The shaping of a wide, shallow channel can be done in many ways. Where heavy power machinery is available, the best method is to use a motor grader working along the axis of the channel, throwing the earth outward toward both sides of the channel until the desired shape is obtained. It will then be necessary to spread the spoilbank along the adjoining fields. Lacking a motor grader, a simple drawn grader may be used.

It is also possible to use less efficient equipment. For example, the channel may be shaped with a bulldozer working laterally from one side, and moving upstream somewhat with each pass and reverse motion. If the soil is not too heavy or dry, and the moisture content gives good workability, the channel may be shaped with the blade of a farm tractor, or with a slipscraper, pulled either by tractor or animals. Finally, under conditions of surplus labor and scarcity of machinery and fuel, as in many of the Third World countries, the waterway channel may be shaped by hand.

The establishment of a grass cover on the waterway under arid or semiarid conditions is not always easy. Under conditions of irrigation, however, the solution is relatively simple.

* A full treatment of the design of open channels is found in Reference 1.

The grass is seeded during the dry season, when no runoff is expected, and germinated with irrigation. If irrigation is not available, the grass must be established during the rainy season, when there is danger that the channel flow will wash the seeds away before they germinate and become established. There are several ways to solve this problem. The best is to establish the grass cover during the rainy season, while preventing the flow of any substantial amount of water until the second year. If the watershed is intended to drain a terrace system, the terraces can be built only after the grass cover is completed. However, field conditions do not always permit this.

An alternative method is to develop the grass cover by sodding, rather than by seeding. An area of suitable grass must be prepared at a grass farm in advance of the operation. After the first rains, blocks of sod, or tufts of growing grass, are transplanted to the channel surface at a spacing which will ensure fairly rapid filling in during the season, to obtain a good grass cover in 1 year. Some grasses are better than others for this purpose, especially if they are spread by rhizomes.

A third method is to plant the seed at the beginning of the rainy season, and to cover the surface with some form of mulch, such as straw, leaves, or other vegetative residues. If the slope of the channel is large, peg the mulch down with sticks and rope. Once the grass is well started, the mulch can be discarded so that it will not interfere with the flow. Other methods may be developed according to the local conditions of climate and types of grasses available.

Another method of establishing a permanent grass cover on the waterway is to seed it first with a quick-growing annual grain crop, to get earlier protection from the first rains. The perennial grass can then be planted by rhizomes or tufts between the rows of grain. It is best to cut the grain crop early, before seeds are formed, and let it lie on the surface as a mulch to protect the young perennial grass.

A word should be added regarding the maintenance of grassed waterways. If neglected, they can become a patch of weeds and trash which will not allow the passage of the design flows. The grass cover should be clipped or mown regularly to keep the channel clear and encourage a better cover. It is possible to graze the waterway at certain times of the year, if the ground is not too wet, provided that adjacent fields are protected, either by fencing or a watchful shepherd. If the waterways are wide, and take up considerable field area, the hay or grazing taken off the channel surface will be an economic benefit which can help to partially pay for the waterway.

III. DIVERSION DITCHES

A diversion ditch is a channel that is generally built across the slope to divert surface flow from its natural course to another outlet. Diversion ditches may serve the following purposes:

1. Where a cultivated field of moderate slope lies at the foot of a steeper slope or hill which is not cultivated, the first soil conservation measure is to isolate the field to be treated from the larger watershed lying above it. This can be done by building a diversion ditch along the upper border of the field. The water thus diverted should be discharged into some protected waterway or other safe outlet. The measures subsequently applied to the field, such as terracing, etc. will have to handle only the water from rain falling on the field itself. This is the principal reason for installing a diversion ditch.

2. In the case of a deep gulley which is being eroded "headward" by an overfall at its upper end, a diversion ditch may be built around the eroding gulley head before other control measures are attempted. This will be treated more fully in Chapter 9.

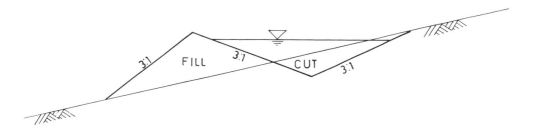

FIGURE 7. Typical cross section of a diversion channel.

3.   If a spring or other form of ground water breaks out at the surface of a sloping field, it may be captured and led off to a prepared outlet by means of a diversion ditch. Such a ditch may act as an interceptor of ground water seepage close to the surface, along its entire length. This condition, however, is more frequently found in humid rather than the semiarid or arid regions.

4.   The diversion ditch may be used to safely carry away a discharge of water from some source point, such as flow from the roofs and paved lots of a built-up area that discharges onto an agricultural field, or the spillway discharge from a small reservoir.

A. Design

The general cross section of a diversion ditch is shown in Figure 7. The flow section lies on the upper side and the excavated material is placed in a compacted embankment on the lower edge. The flow channel section and the embankment are usually shaped together like a sine curve, or, more accurately, like a parabola and an inverted parabola. The hydraulic design has the following steps:

1.   The design flow for waterways can be found using the method described above. However, there is an important difference with respect to the flood frequency or return period to be selected. As stated, this factor depends upon the hazard or damage which might be incurred in the event of failure due to underdesign. If a waterway is over-topped, it will, in most cases, cause minimum damage to the adjoining fields, since the channel lies in the lowest part of the field. Consequently, a short return period may be taken, such as the rainfall or flood which may occur on an average of once in 2 years. However, a diversion ditch takes the water out of its natural course. If it should be overtopped because of inadequate design, considerable damage in the form of gullies and washouts may be caused to the field below. This justifies the use of a return period of once in 5 or 10 years. The actual decision is not automatic but depends upon an estimate of the hazard and the judgment of the engineer.

2.   The hydraulic design of the channel section follows the normal procedure using Manning's formula. In most cases the diversion ditch will eventually be covered with a permanent grass cover. The retardance factor, n, will be somewhere between 0.4 and 0.7. The longitudinal slope may vary, but it should be somewhere in the neighborhood of 1% to obtain safe velocities.

3.   In general, a cross section should be chosen that is not too wide, so as to minimize the encroachment on the field. The depth will generally be much greater than for waterways, as the design discharge will probably be greater. The hill area above a diversion may be large, and the runoff factor on the steeper land may also be high. Therefore, diversions must be able to carry large discharges at higher velocities. For this reason that diversions should almost always be under a permanent low-growing grass cover.

B. Construction

Because diversion ditch sections are often deeper and narrower than waterways, some difficulty may be encountered in constructing them by motor grader, unless the blade can be tipped to a fairly high vertical angle. Modern motor graders, however, can usually do this, especially if the blade can be raised at one end while offset. The older type of drawn grader used for road patrolling may not be adequate. If the land along the upper margin of the ditch is not too steep, the channel and the dike can be formed with a bulldozer or front loader working laterally. The loose embankment must then be shaped with a tractor blade working longitudinally, and compacted. The establishment of the grass cover is best done by hand planting seedlings, roots, or blocks of sod.

C. Maintenance

Diversion ditches commonly suffer from very poor maintenance. Unlike waterways, they are not in the cultivated part of the field, and are neglected and allowed to turn into strips of wild weeds. It is worthwhile to keep the diversions clear for better flow conditions during the flood season, and to keep nearby fields free of weeds.

REFERENCES

1. **Finkel, H. J.,** *Handbook of Irrigation Technology,* Vols. 1 and 2, CRC Press, Boca Raton, Fla., 1982.
2. **Little, W. C., Piest, R. F., and Robinson, A. R.,** SAE research program for channel stability and gulley control, *Trans. ASAE,* 362, 1980.
3. **Nathan, K.,** An improved design procedure for grassed waterways, *Trans. ASAE,* 66, 1972.
4. **Ree, W. O. and Palmer, V. J.,** Flow of Water in Channels Protected by Vegetative Linings, Tech. Bull. No. 976, U.S. Department of Agriculture, Washington, D.C., 1949.

Chapter 7

ENGINEERING MEASURES FOR SOIL AND WATER CONSERVATION: TERRACING AND BENCHING

Herman J. Finkel

I. TERRACING

Terracing, as discussed in this section, refers to the broad-based terrace constructed in the form of a ridge and channel, the entire surface of which is cultivated as part of the field. The ancient system of "staircase" agriculture, now known as benching, will be treated separately.

The first broad-based terraces were developed in the humid eastern part of the U.S. Their purpose is to shorten the length of the slopes for reduction of soil erosion, and to lead off surplus storm water at a safe velocity to a protected outlet, usually a grassed waterway. These terraces have a cross section which is wide and flat enough to permit the use of farm machinery both on the ridge and in the channel. The alignment is more or less along the contour, with a small, variable slope which increases toward the outlet to accomodate an accumulation of surface flow. Where the contour has sharp bends, as across a draw or depression, the land is first filled to allow for a smoother curve of the alignment. The cross section is, consequently, uniform along the entire terrace, and can be built by successive longitudinal passes back and forth along the length of the terrace. The terraces are laid out at a constant vertical interval, and because of topographic changes, the width between the terraces is rarely uniform. This lack of parallelity creates problems of point rows and/or correction areas which are inconvenient for farming, especially with large machinery. Many of the terraces installed in the 1930s and 1940s were neglected and abandoned in later years. It became clear to the conservation planners that the extra cost and effort to obtain a system of parallel terraces would be justified.

The "drainage type" terrace described above is not suitable for semiarid regions. Water is too precious a resource to be disposed of, and every effort must be made to retain it on the field, while preventing both local waterlogging and soil erosion. This is accomplished by the "absorption type" terrace.

The most important absorption type of parallel terrace for the semiarid regions is the flat channel terrace, or, as it is sometimes known, the Zingg conservation bench terrace. A typical cross section of this terrace is shown in Figure 1. The bed of the channel is level in the transverse direction and the water is retained by a low ridge. The width of the channel bed depends upon the slope of the land, the allowable depth of cut to subsoil, and the width of the machinery which will be used to construct the terrace. If subsequent agricultural operations will be done with large power equipment, the channel bed should accomodate the widest machine, usually the grain drill or the combine. The channel grade should be as small as possible consistent with the following principles: (1) maximum conservation of rain water with minimum discharge to the outlet, and (2) ponded water should not remain in the channel for longer than 48 hr. If the soil has a high infiltration rate the channel grade may be zero. The value of this infiltration rate will be considered presently.

The vertical interval, VI, of this type of terrace in semiarid regions may be estimated by the following formula:

$$VI = 0.8S + 1 \qquad (39)$$

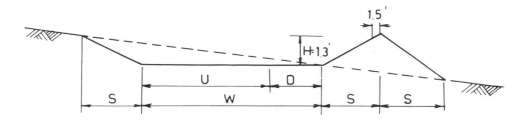

FIGURE 1.   Cross section of the Zingg conservation bench terrace.

where VI is in feet, S is the weighted average of the land slope in percent, and 1 is a constant in feet for areas with full plowing and little or no crop residues.

$$VI = 0.25S + 0.30 \qquad (40)$$

where the units are in meters.

For example, if the slope, S, is 5%, the vertical interval of the terraces would be 5 ft, or 1.55 m. On this field, the average lateral spacing between the terraces would be 100 ft, or about 30 m. The terrace channel should be designed to carry the runoff from a 24-hr, 10-year rainfall frequency. Using, for instance, data from Jerusalem, the rainfall would be 100 mm, or 4 in. On a strip normal to the terrace (1-m wide), the drainage area above the terrace would be 30 m². With an assumed coefficient of runoff of, for example, 15%, the total volume of runoff would be 0.45 m³/m for the length of the terrace. If the channel bottom width is 3.0 m, the depth of the collected water would be 0.15 m, or 150 mm. This value divided by 48 hr would give about 3 mm/hr. Consequently, if the infiltration rate of the soil in the channel bed is at least 3 mm/hr, all of the estimated runoff can be absorbed, and it is not necessary to give the channel a longitudinal slope.

Expressed in FPS units, the rainfall of 4 in. on a strip 1-ft wide and 100-ft long would have a volume of 33 ft³. Multiplied by the runoff coefficient of 15%, this would equal 5 ft³/ft for the length of the terrace. If the channel width is 10 ft, the depth of the collected water would be 0.5 ft, or 6 in. Over a period of 48 hr, the average infiltration rate would be 0.12 in./hr. If the soil has at least this rate, the channel bed may be built without a slope for drainage.

In this example, rather high rainfall values and an intermediate value for the runoff coefficient were assumed. The infiltration rate is not the initial rate, which could be quite high, but is the steady rate after several hours, which would be considerably lower. If the terrace system is installed in a field which has a relatively shallow soil over rock or hardpan, the infiltration rate will not govern; it will be superceded by the capacity of the soil profile to store gravitational water, i.e., between the field capacity and the saturation percentage. In the case of a soil with lower capacity to absorb water, or if the design rainfall and/or the coefficient of runoff are higher than in the above example, a small longitudinal slope should be provided which does not exceed 0.2%. At the end of a terrace with a channel grade, provision must be made for the excess water to drain away, either to a grassed waterway or to some other type of outlet.

In the previous example the depth of the water in the channel would be 15 cm, plus a certain allowance of approximately 10 cm for freeboard, after settling of the ridge. Neglecting the infiltration which would take place during the first 24 hr of the design rainfall, a value of 25 to 30 cm of channel depth is required. From this, it is possible to determine the depth of cut along the upper margin of the channel. It is also a function of the width of the channel and the slope of the land. With a shallow topsoil on a high slope, the channel width must be restricted to avoid cutting down to the subsoil. For a deep topsoil on a small cross slope,

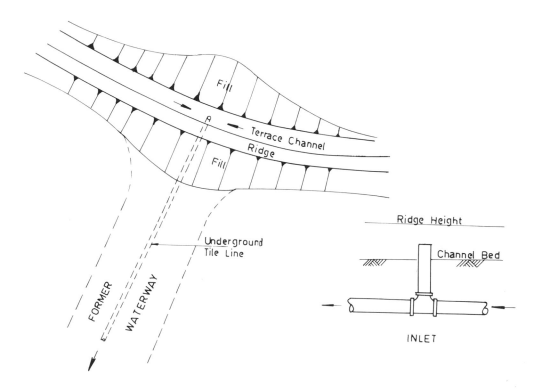

FIGURE 2. Underground tile outlet for terraces.

the channel width can be correspondingly wider. It is easiest to work out the limits of depth and width graphically by drawing the cross section and surface slope to any convenient scale, as shown in Figure 2.

In fields where the slope is larger, and the required width of a grassed waterway would occupy an excessive area, the modern trend is to discharge the outflow from a terrace system into an underground tile outlet, as shown in Figure 2. This permits the cultivation of the entire field with no interruptions. The intake to the tile line should be placed somewhat above the bottom grade line of the channel, and should be provided with an orifice plate which can control the outflow so that only the surplus which the channel bed cannot contain and absorb in 48 hr will be allowed to pass to the tile outlet. Since the tile line will generally be located in the draw or low part of the field, the terrace at this point will be built on fill to keep it level. Extra fill material may have to be brought in from a borrow pit, but, in order to save this expense, it is possible to design the terrace alignment in such a way as to balance cut and fill along its entire length. In order to accomplish this on irregular topography, it may be necessary to vary the cross section of the terrace at different reaches. The design procedure to achieve this balance is somewhat tedious, but has been solved by means of computer programs.[6]

The direction of plowing must be parallel to the terrace alignment, and the backfurrow should fall on the ridge, to provide a buildup of the hydraulic section. If, after several years, however, the ridge becomes too high and the channel too deep, the plowing pattern can be reversed, with the deadfurrow on the ridge. In rolling or irregular topography, the use of a fixed vertical interval may result in considerable variation in the horizontal spacing between a pair of terraces which is too great to correct by changing the cross section along the route and trying to balance the cut and fill. In this case, parallelity msut be sacrificed and point rows are inevitable. A pattern for locating the point rows may be similar to that shown for contour cultivation in Figure 5 of Chapter 4.

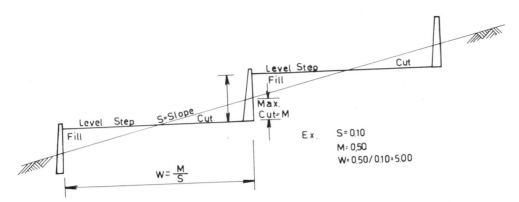

FIGURE 3.   Geometry of level benches.

There are several variations on the broad-based terrace in addition to the Zingg type, but for semiarid regions where water conservation is of paramount importance, it is recommended that the level, closed type should be used wherever possible. In addition to water conservation, the level, impoundment-type terrace retains more soil than any of the other types. However, when such terraces are installed on an erosive field, the channels will quickly become filled with trapped erosion sediment; consequently, they should be cleaned and maintained regularly. The soil removed from the terrace channel should be returned to the field.

II. BENCHING

Benching, or bench terracing, refers to the practice of converting a sloping field into a series of almost level areas or steps. It has also been called "staircase farming". Benching is one of the oldest soil conservation systems in history. It was practiced extensively by the Chinese and Indonesians thousands of years ago, and was developed independently, in a spectacular way, by the pre-Columbian cultures of Peru and Mexico.[4] In the Middle East, it is said to have been first developed by the Phoenicians along the western coast of the Mediterranean and then carried by them to southern Europe and northern Africa.

A. Design

The essential features of benching are shown in Figure 3. The step may be of varying width in inverse ratio to the degree of slope of the land. Another factor affecting the optimum width of the bench is the depth to which the soil may be cut along the upper margin of the bench. Thus, for example, on a 10% slope, if the maximum allowable cut is 0.50 m, the width of the bench would be 10 m. On a 20% slope, where the maximum allowable depth of cut is 0.30 m, the width of the bench should be 3 m.

In cross section, the surface of the step may be level, but this is not always necessary; a small slope toward the retaining wall may be tolerated. This reduces the height of the walls, and hence makes the entire project less expensive. The longitudinal slope of the bench is usually level, but similarly, a slope from 0.5 to 1.0% may also be tolerated. This sometimes simplifies the layout and construction, by fitting in the benches between layered outcrops of bedrock, which may not be aligned with the true contour, as is typical of the Judean Mountains around Jerusalem.

Another consideration affecting the slopes of the ground surface behind the wall is the expected amount of runoff. If it is estimated that all of the design rain can be captured and infiltrated into the land on the step, then the bench can be made level. If, on the other hand,

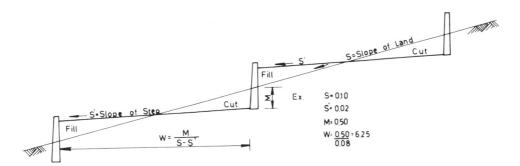

FIGURE 4. Benches with forward lateral slope.

the design flood will produce runoff beyond that of the terrace step to absorb, then the small longitudinal slope can carry the surplus to some prepared outlet at one end of the terrace. This is a problem of balancing the probabilities. If a rain less than that producing the design flood falls, we would like to absorb as much as possible. The longitudinal slope in such a case should be minimal. The effect of the transverse slope in draining away excess runoff is minimized if the crest of the wall is somewhat higher than the land surface.

The wall dividing the benches will have a height of twice the maximum depth of cut, M, as can be seen in Figure 3. In addition, there should be a certain amount of freeboard (10 to 20 cm) above the final surface level. The effect of the lateral slope of the land between the walls is shown in Figure 4. It will not reduce the height of an individual wall, which, for the typical shallow soils found on steep slopes, is twice the permissible depth of cut. However, will increase the spacing of the walls, W, for the same allowable cut, M, and reduce the total amount of walls required in a given field.

If the depth of cut is not a limiting factor, the designer has the choice between making fewer, higher walls, or a greater number of lower walls. Generally speaking, the height and spacing of the walls should, in such a case, be defined by the minimum width of the bench convenient for agricultural operations. If the bench will be plowed and planted to grain or a row crop, the minimum width will be governed by the machinery, such as grain drill or harvesting equipment, which will be used. If the grain will be planted and harvested by hand, there is no limiting minimum width for the bench. If the benches are to be used for tree crops, with one row of trees to a bench, the minimum width may also be quite small. Aside from these considerations, however, it is generally true that more, lower walls are easier to install and more economical than fewer, high walls.

In laying out a field for bench terraces, provision must be made for access to each bench. If the work will be done by hand labor, some type of stairway should be built in the center or along the sides of the field. If the work will be done with tractors, earthen ramps should be provided at the ends of the benches to permit the tractor to move from one bench to the next.

B. Construction

The wall may be constructed in several ways. On moderate cross slopes, with low vertical intervals between relatively wide benches, the wall may be made of earth at a slope of about 1 horizontal to 4 vertical. The face of this steep slope can be stabilized with some sort of permanent vegetative cover, such as a perennial grass, providing it is not aggressive. Succulents, such as *mesambrianthimum,* can be planted to hold the slope, but these require a certain amount of moisture. The ideal material for stabilization, where there is a dearth of stone, would be some sort of dry, inert crop residue which can be staked or otherwise attached to the steep slope.

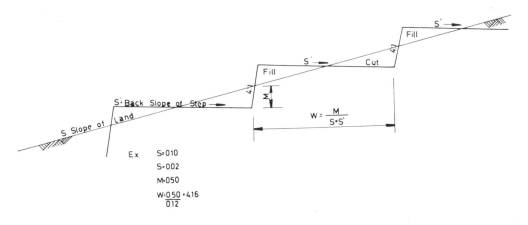

FIGURE 5.   Benches with reverse lateral slope.

With an earth embankment supporting the lower edge of the bench terrace, there is usually no freeboard to prevent runoff from spilling over into the next lower step. In such a case, it is possible to introduce a lateral, reverse slope on the bench, so that the excess water will drain away from the top of the lower wall, and flow in a natural channel in the cut portion of the section, where the bench surface meets the toe of the next wall above it. This prevents loss of runoff water, but is applicable only where the soil is deep enough to permit the extra cut required by the reverse slope (see Figure 5).

In areas where field stone is available, the wall can be built of dry rubble masonry. The lower (exposed) face of the wall should have a slight batter (1 on 5) inward at the top, to give it more stability. The taller the wall, the more the degree of batter required. It requires skill and experience to build up good terrace walls with rough field stone, and to ensure that they are properly anchored in the ground. The final result, when done correctly, is beautiful and long-lasting, but the cost is relatively high. It can be justified in very hilly regions where there is a dearth of good land with moderate slopes, and where there is a surplus of seasonal labor with no alternative occupation.

C. Agricultural Operations

When bench terraces are used for orchards, and the width allows only a single row of trees on each, the question arises as to where to plant the trees on the cross section. From the point of view of agricultural operations, such as pruning, spraying, and picking the fruit, it is best to plant the trees in the middle of the step. However, from the point of view of soil fertility and root development, it is sometimes thought that the trees should be planted along the outer edge of the step, since the soil there is topsoil fill, and much more fertile than the cut along the foot of the next higher wall. The agricultural operations mentioned earlier can be handled from ladders resting on the next lower step, if the vertical interval between the steps is not too great.

The effect of bench terracing on the rate of soil loss is very great. If the benches are almost level, and the intervening walls are made of stone, the soil loss will be practically zero, unless an unusual storm exceeds the capacity of the benches to hold water behind the freeboard of the wall, and overtopping occurs. If the intervening wall is made of sloped earth, the amount of soil loss will depend upon the degree of stabilization of this slope. If a close-growing vine is used, there will be almost no soil loss, but the resulting loss of moisture could be serious in semiarid regions.

REFERENCES

1. American Society Agricultural Engineering, Design, layout, construction and maintenance of terrace systems, *ASAE Yearb.,* 530, 1978.
2. **Black, A. L.,** Conservation bench terraces in Montana, *Trans. ASAE,* 393, 1968.
3. **Buchta, H. G., Broberg, D. E., and Liggett, F. E.,** Flat channel terraces, *Trans. ASAE,* 571, 1969.
4. **Cook, O. F.,** Staircase farms of the ancients, *Natl. Geogr. Mag.,* 474, May 1916.
5. **Dominy, P. F. and Worley, L. D.,** Design and construction techniques for parallel terraces, *Trans. ASAE,* 580, 1966.
6. **Forsythe, P. and Pasley, R. M.,** Terrace computer program: practical applications, *Trans. ASAE,* 512, 1969.
7. **Haas, H. J. and Willis, W. O.,** Conservation bench terraces in North Dakota, *Trans. ASAE,* 396 and 402, 1968.
8. **Hauser, V. L. and Cox, M. B.,** Evaluation of Zingg conservation bench terraces, *Agric. Eng.,* August, 462, 1962.
9. **Hauser, V. L., Van Doren, C. E., and Robins, J. S.,** A comparison of level and graded terraces in the Southern High Plains, *Trans. ASAE,* 75, 1962.
10. **Larson, C. L.,** Geometry of broad-based and grassed-backslope terrace cross sections, *Trans. ASAE,* 509, 1969.
11. **Mickelson, R. H.,** Conservation bench terraces in Eastern Colorado, *Trans. ASAE,* 389, 1968.
12. **Phillips, R. I.,** Tile outlet terraces: history and development, *Trans. ASAE,* 517, 1969.
13. **Rochester, E. W. and Busch, C. D.,** Storage and discharge requirements for impoundment terraces, ASAE Pap. No. 74-2034, American Society Agricultural Engineers, St. Joseph, Mich., 1974.
14. **Thomas, D. B., Barber, R. G., and Moore, T. R.,** Terracing of cropland in low rainfall areas of Kenya, *J. Agric. Eng. Res.,* 25, 57, 1980.
15. **Wittmuss, H. D.,** Topographic modification of land for moisture entrapment, *Trans. ASAE,* 384, 1968.
16. **Wittmuss, H.,** Construction requirements and cost analysis of grassed-backslope terrace systems, *Trans. ASAE,* 970, 1973.

Chapter 8

ENGINEERING MEASURES: WATER HARVESTING

Herman J. Finkel and Moshe Finkel

I. INTRODUCTION

Water harvesting is the collection of runoff and its use for the irrigation of crops, pastures, and trees, and for domestic and livestock consumption. Since this definition could cover almost all fields of water resource development, it must be added that the term "water harvesting" is usually used for the development of marginal waters in arid or semiarid regions. The projects are generally local and of a small scale that do not include the treatment of water or its conveyance over long distances. They may be classified according to the source of the water and the use to which it will be put. The latter has two categories: irrigation and for human and/or animal consumption. In the case of water for consumption, two additional elements must be included: storage and treatment. The storage is needed because the rate of consumption of water will generally be slower than the rate of harvesting. In the case of water for irrigation, the soil itself serves as a reservoir for a certain period of time. For a longer storage period seasonal reservoirs may be required. Dams on intermittent streams to contain flash floods are also considered to be a form of water harvesting. However, this chapter will deal only with water harvesting techniques for irrigation purposes, and only short-term storage facilities.

II. DESIGN PRINCIPLES

A. The Design Model

The design principles of a water harvesting project are similar to those of other hydraulic projects requiring a wide range of input. However, many water harvesting projects are built in remote areas of Third World countries where basic data may be scant. They will usually be installed by the local rural population without expert guidance, and using hand labor instead of heavy machinery. Under these circumstances it will be almost impossible to calculate and develop exact hydrological models, or to define design inputs such as flood discharges, volumes, durations, and peaks.

In many regions local rules of thumb are used for designing the structures. For the hydrological design a more or less universal criterion is used which is the ratio of the catchment area to the cultivated area. Where this ratio is known or assumed, the possible size of the field to be irrigated can be easily determined. The size of the catchment area can be either surveyed or estimated in the field or measured on topographic maps, if such are available. Where a planimeter is not available an approximation can be made by counting squares. Once established, this ratio becomes the basic design tool.

In many parts of the world the accepted rules of thumb for the ratio vary from 1:5 to 1:40, depending upon rainfall depth and distribution, watershed characteristics, runoff coefficients, and the water requirements of the crops to be irrigated. The following procedure may be helpful in deriving a rule of thumb ratio.

$$\text{Catchment area} \times \text{Design rainfall} \times \text{Runoff coefficient} \times \text{Efficiency factor} = \text{Cultivated area} \times [\text{Crop water requirement} - \text{Design rainfall}] \quad (41)$$

Therefore, the ratio equals

$$\text{Ratio} = \frac{\text{Catchment area}}{\text{Cultivated area}} = \frac{\text{Crop water requirements} - \text{Design rainfall}}{\text{Design rainfall} \times \text{Runoff coefficient} \times \text{Efficiency}} \quad (42)$$

B. Crop Water Requirements (CWR)

CWRs are a function of the specific crops to be grown and the climatic conditions. Estimates may be developed from pan evaporation data and known coefficients for different crops. Studies can be made to the level of accuracy required, including the calculation from the climatic records of probabilities of occurrence of different rates of evaporation. Where evaporation data is not available, estimates may be based upon experience or data from areas of similar climatic conditions. (A fuller treatment of crop factors for use with pan evaporation data may be found in *Handbook of Irrigation Technology*, CRC Press, 1982.)

C. Design Rainfall (DR)

DR is the rainfall volume for which the project is designed. It is the amount of rainfall which, from experience, may be expected to fall during one rainy season, with a probability of occurrence of 33%. If the rainfall in a given season is less than the design rainfall, the water supply will be inadequate. If more, the surplus would require drainage to prevent endangering the crop. Either extreme threatens the success of the project. The distribution of this design volume of rain must also be taken into consideration. If it all falls at once, facilities must be provided to drain away the surplus, preferably to another crop area.

D. Runnoff Coefficient (RC)

RC is the percent of the rain water which actually flows down the slopes as surface runoff. This varies with the degree of slope, soil type and geology, vegetative cover, antecedent rain, rainfall intensity, and return period. This coefficient may be derived from rainfall and stream gauging records, if available. It may also be measured in small test plots. However, in the remote field areas it will be estimated by the designer from his experience (see Chapter 2 for a fuller discussion of this coefficient).

E. Efficiency Factor (EF)

EF is a factor taking into account the difference between the rainfall patterns and the rate of water consumption by the crop. When these do not fully correspond, the factor is low. When the rain falls in well-distributed, gentle showers, the factor will be high. EF may range from 0.25 to 0.85, and must be selected according to the discretion of the field engineer, based upon his observations and experience.

Note that the design rainfall must be subtracted from the crop water requirements as this rainfall is assumed to fall over all of the cultivated area, as well as on the catchment above it.

For example, for Lokichoggio in Turkana, Kenya: crop — sorghum or maize; water requirement — 575 mm or more, with a 67% probability (2 out of 3 years); average annual rainfall — 522 mm; design rainfall: 380 mm (equal to or less than) for 33% probability, or 1 in 3 years; runoff coefficient — 13%. (Note that this is the yearly value based upon daily rainfall data for a record of 2 years, assuming a range of factors for various size rains. The daily values are based upon data from nearby test plots.)

$$\text{Ratio} = \frac{\text{CWR} - \text{DR}}{\text{DR} \times \text{RC} \times \text{EF}} = \frac{575 - 380}{380 \times 0.13 \times 0.5} = 7.89 \quad (43)$$

The ratio to be used is 1:8. To irrigate 1 ha of cropland, runoff from 8 ha must be collected.

III. RAIN WATER HARVESTING TECHNIQUES

A. Contour Ridges and Contour Furrows

This technique consists of small earth bunds with heights up to 0.40 m, which are built on the contour. The earth for forming the bunds is taken from along the upper side of the bund creating a furrow along the bund. The distance between the ridges may vary from 2 to 20 m depending upon the catchment/cultivated ratio for the region and the type of crop to be grown. The space between a pair of adjacent ridges is the catchment area, while the area of the furrow itself is the cultivated area.

This method is applicable on slopes of 1 to 3% having soil of at least 1-m deep. The rainfall should be at least 200 mm/year, but the method has also been successfully applied in areas with an annual rainfall of only 100 mm.

Bund dimensions should be as follows: base width of at least 1.0 m, crest width of at least 0.20 m, bund height of a maximum of 0.40 m, distance between bunds of at least 2.0 m.

Excavating the furrow and throwing up the bund may be done by mechanical means such as a plow, a disk bunder, either tractor drawn or with draft animals, or by a motor grader. Lacking these, they may also be built by hand. The layout of a field may be simple, requiring only a hand level to find the contour line, and a tape to measure the horizontal distance.

In order to ensure an adequate supply of water to the furrows the area between them should be weeded and kept clear. The soil may be compacted in order to increase the runoff factor and thus harvest more water. If this will be done, it must be taken into account at the design stage when the distance between consecutive bunds is determined.

The bund should be compacted during construction in order to improve chances of retaining the water accumulated behind it. The furrow should be wide and shallow, with the crop planted along both sides of the furrow on its side slopes rather than along the bottom. The planting area should be cultivated in anticipation of the rainy season. Seeds should be sown immediately after the first rains. This technique has a very small safety factor because of two basic characteristics: the narrow furrow allows for a very small storage capacity and the low ridge adds to the risk of bund breakage in years of higher rainfall.

The system is recommended only for areas with a known regular rainfall pattern. It is used extensively in many semiarid regions where water harvesting has become a tradition. Over many centuries of trial and error the farmers have found the exact dimensions for achieving maximum results. In areas where the rainfall patterns are not yet fully known or in areas where the rainfall pattern has extreme fluctuations, as in Turkana, this method would not be recommended.

B. Semicircular and Triangular Microcatchments

These techniques consist of small bunds shaped either as semicircles or as triangles. They are constructed with the tips of the semicircles or triangles on the contour line (Figure 1). Water is impounded behind the bunds to the level of the contour and eventually overflows, spreading to the next lower tier of bunds. These structures, also termed microcatchments, are constructed in long rows along the contour lines. The radius of the semicircle and the dimensions of the triangle are determined by the catchment area/cultivated area ratio, as is the distance between the rows. The structures in the consecutive rows are staggered so that the catchment in a lower tier can easily collect the overflow from above. The bunds are 0.30 m in height with a base width of at least 0.80 m, side slopes of 1:1.5, and a crest width of 0.20 m.

Semicircular hoops have radii ranging from 4 to 12 m and are used mainly for improvement of grazing areas. The results may be quite dramatic as were seen on the test plots in the Baringo area of Kenya. A few handfuls of seed were spread prior to the rains. In the first

FIGURE 1.   Semicircular and triangular bunds.

season the grass grew only along the rim of the bund. By the second season the grass had spread throughout the areas enclosed by the semicircles. Seeds used included *Eragrotis superba* and *Cenchrus ciliaris*.

Triangular bunds are used mainly for tree cultivation. This system was developed in the Israeli desert and are also known as ''Negarim''. Two trees are usually planted at the lower apex of the triangle. In some cases a large hole is dug near the tree which serves as a reservoir for as much as 1000 ℓ of water. The system was very successful in a region with 300 mm of annual rainfall or less.

Microcatchments of both types may be used in areas with rainfall as low as 150 to 200 mm per season. They may be easily laid out in the field with hand levels, and constructed by hand labor. In North Kenya it was found that 4 man-hr were required to build triangular microcatchments of 10 × 10 m in length. Semicircular hoops are recommended as a quick and easy method to construct for improving pasture areas in semiarid regions suffering from drought. They are especially suitable for areas where farming is not socially acceptable to the nomadic pastoralists.

C. Trapezoidal Bunds

This technique is similar in principle to the semicircular microcatchments, but is used for enclosing larger areas and impounding larger amounts of water. They consist of a trapezoidal shaped bund with its tips lying on the contour line (see Figure 2). The bund may enclose

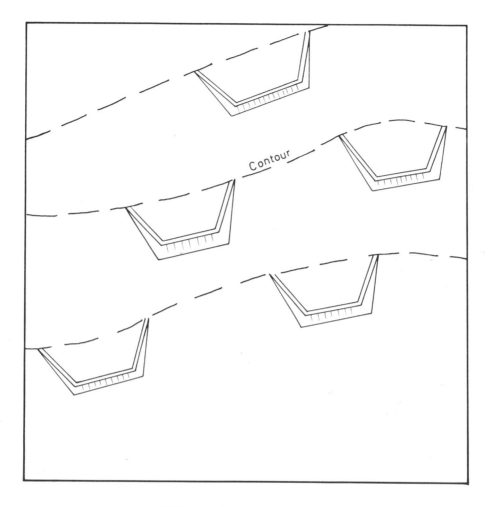

FIGURE 2.   Trapezoidal bunds.

an area of 0.25 to 2 ha. The size depends upon the catchment/cultivated area ratio which also defines the distance between the bunds. The area enclosed by the bund is used for cultivation, whereas the area between the rows of bunds is the contributing catchment. Water is impounded to the elevation of the contour line connecting the bund tips. A freeboard of 0.40 m is recommended. An option is to protect the tips where the water overflows by stone paving or to construct simple stone spillways. Where large water quantities are expected, canals may be constructed to handle the overflow from one tier of bunds to the next. Spillways paved with stone may be constructed at the tips to protect against erosion. Another variant includes land leveling within the bund to achieve uniform irrigation throughout the cultivated area. This is not recommended where the limited fertility of the topsoil may be lost.

Trapezoidal bunds are used for planting either field crops or trees. The layout of an area includes a tier of trapezoidal bunds along a contour line. The row along the next lower contour line is staggered as with the semicircular hoops allowing the overflow from the first tier to fill the bunds below. The staking out is done with hand levels, prisms, and measuring tapes. The construction can be done either by mechanical equipment or by hand labor. The slope of the land selected for bunding should not be steeper than 3% to avoid high bunds or small fields. The breaking up of an area into rows of trapezoidal bunds will have less risk of damage due to failure than the same area enclosed by a single large bund.

IV. WATER HARVESTING FROM SEASONAL RIVERS

A. General

The diversion of seasonal rivers usually requires larger and more ambitious schemes than water harvesting of sheet flow, as it makes use of concentrated discharges already flowing in the stream bed. The basic design ratio of catchment/cultivated area applies here as well. However, if stream flows can be estimated, or if records are available, they would provide a better basis for design than the multiplication of the design rainfall by an estimated coefficient of runoff. The estimated probability or return period to be used would be about once in 2 or 3 years for calculating minimum supply, and possibly once in 10 or 15 years for calculating necessary spillway protection. There are several types of structures used for water harvesting from intermittent streams. They will be described briefly.

B. Diversion to Terraces

Small weirs are built of earth, rock, or gabions across the streams. The flood water is diverted and guided through a series of basins where crops are cultivated. The area enclosed within the basins is the cultivated area and the watershed above the weir is the contributing area, only when all of the water is diverted. Surplus flow will spill over the weir. A well-planned project includes overflow protection at the basins so that varying flow from the river can be accomodated. If the fields to be irrigated lie at some distance from the river, a diversion canal must be constructed. Establishing the crest height of the weir is a matter of some concern. If it is too low, a larger proportion of the flow will be wasted. If it is too high, more water may be diverted than can be accomodated by the flooding basins. In semiarid regions it is safer to have the crest higher, with provision for disposing of the surplus, should it occur, at the outlet of the lowest basins.

Another critical problem is the silt carried by the river which may settle behind the weir and in the diversion canal, eventually blocking it completely. It is therefore necessary to build silt traps upstream which must be periodically cleaned out. If this is not feasible, frequent maintenance of the diversion works will be required to keep them functioning.

A variation of this system is to build the diversion weir of earth and brush, which is considered expendible and can be rebuilt before every flood season. This approach seems simple but it provides the farmer with less control of the flow, as it is impossible to predict when a major flood will occur and wash away the weir.

The flow diverted to the diversion channel may be distributed to the leveled basins mainly in two ways. The first is to have turnout structures along the channel which direct the entire flow to each adjoining basin in turn. These basins may be on one or both sides of the diversion channel, according to the topographic conditions. The gates can all be set before the flood is expected, and the highest one or two on the line should be operated first. When those basins are filled the water can be turned into the next lower basin, and so forth until the lowest basin is irrigated. Any surplus flow can then be carried back to the main stream by a drainage channel. The disadvantage of this system is that it requires manual operation of the gates, which may be very inconvenient if the flash flood should occur at night. Automatic gates with timers are available, but these are usually beyond the means of the typical arid-zone population in Third World countries.

The second principal method is to run the water consecutively through all the basins. This is commonly used in Pakistan and has been adapted for use in other parts of the world. It consists of basins enclosed by a bund built on a slight gradient. Water reaching the end of the bund overflows to the next basin where it flows along the bund, graded in the opposite direction. This creates a very long path for the flowing water, permitting maximum infiltration as in gravity irrigation systems. The advantage of this system is that it does not require any manual operation. The disadvantage is that the application of water is not very uniform, as the upper basins get far more than the lower.

C. Retardance Dams

Retardance dams (sometimes called check dams) are small stone or gabion structures which are built across seasonal rivers to retard or check the flood flows. The crest of the structure acts as a spillway, allowing peak discharges to flow over the dam and continue downstream. Check dams extend the duration of the flow, having a flood-routing effect on the hydrograph. This enables a larger quantity of water to infiltrate into the alluvial stream bed. It is a form of aquifer recharge. The water can be harvested through existing nearby wells, or from water holes drilled in the stream bed. The aquifer may actually be a perched, shallow layer of river bed alluvium, or it may extend down to a deeper, wider water-bearing strata. This technique is more effective when a series of retardance dams are built along the same river. It is used on rivers with catchment areas of from 20 km² for local use to large basins of 4000 km² for regional aquifer recharge. Check dams are applicable mainly in areas where the available knowledge includes both hydrological and hydrogeological inputs, and where pumps and an energy supply are available. Likewise, a technical staff must be capable of operating and maintaining the system.

The idea of check dams has been adopted as the policy for harvesting flash floods in the Arava Valley of Israel. Instead of trying to construct large dams and reservoirs at the lower ends of the rivers, small check dams are planned in the upper reaches for the aquifer recharge. A few pilot check dams have already been installed on streams with catchments of 2000 km². Pumping tests in existing wells downstream have shown improved output.

D. Underground Dams

A variation of the check dam is the so-called "underground dam", which is a vertical barrier constructed across the bed of a stream below the ground surface. It is applicable where the valley walls below the surface are of a relatively impervious rock, and the bedrock under the stream bed is not too deep. This geological structure, together with the artificial underground barrier, form a subterranean reservoir in the alluvial fill. If the longitudinal gradient is suitable a series of such underground reservoirs can be spaced along the stream to catch part of the peak flow and much of the tail water. This system has many advantages over the above-ground reservoir, including less evaporation loss, no siltation problem, no expensive spillway construction, no flood hazards due to failure, and no land lost. The water is harvested by shallow pumping from the alluvium. There are several engineering techniques for creating the underground vertical barrier, but their details fall beyond the scope of the present volume.*

E. Water Spreading Terraces: Limanim

On intermittent streams which have small watersheds of a few square kilometers, the lower end of the valley may spread out to a wider, alluvial plain. If the slope is gentle and the soil sufficiently deep, it is possible to construct level terraces across the entire valley, which will effectively obliterate the original channel. The water covers the first terrace and is then drained off at the side, through a stone weir or over a low gabion into the field below. The outlet for this will be placed at the opposite end of the terrace, and continue to change sides with alternate terraces. The water is thus "walked" back and forth across the bank of terraces until it is either fully absorbed or discharged back to the stream at the lower end.

The total area of the spreading terraces is the cultivated area, and the watershed area above the lowest terrace is the catchment area. The ratio of these must comply with the

* The system was developed and the mechanical equipment designed and built by Dr. Gedalyahu Manor and Mr. Amnon Neuman of the Faculty of Agricultural Engineering, Technion, Israel Institute of Technology, Haifa. A somewhat similar system has been tried in North Africa, based upon the use of liquid asphalt heated in the field. The special equipment required for keeping the asphalt hot during placement is more cumbersome.

calculated ratio for the region. It is possible, however, to prepare more terraces than indicated by the ratio, if there is space available. The lowest terraces may not always get water, but in the event of an unusual flood no water will be wasted. These tail water terraces should be planted only to annual crops after the first flood, whereas the higher terraces may be used for perennials such as tree crops. The vertical interval between the terraces should follow the same principles as those for bench terraces discussed in Chapter 7.

These water-spreading terraces are also called *limanim,* which is derived from the Greek term for port. They are commonly used throughout the Negev Desert of Israel by the Forestry Department to form little oases of shade for travelers. Ruins of Byzantine cities in the Negev contain extensive remnants of water-spreading terraces which were developed to a high degree of sophistication. In modern times they are built by the Bedouin to grow a little supplementary grain in the years when there is some rainfall.

F. Underground Membranes

A novel method has been developed* to convert sandy soils and tracts of dune sand into agriculturally productive fields in semiarid regions. The principle is to insert an impervious horizontal membrane into the sand over the entire field area. This prevents the rapid percolation of water down to depths below the root zone of the crops. A machine was designed and built to drag an underground blade at a depth of 60 cm without excavating or turning over the soil. Behind this blade a set of nozzles was mounted through which an emulsified asphalt was injected into the horizontal crevice. As the machine moves forward the overburden of soil fell back in place covering the asphalt. When the emulsion "broke", an impervious membrane only a few millimeters thick was formed. The equipment is 2.5-m wide, and capable of installing the membrane very quickly, in slightly overlapping strips. The same machinery, slightly modified, can place polyethylene sheets at the same depth, instead of the emulsified asphalt. Field trials showed that on very sandy soils, the yields of unirrigated potatoes, peanuts, and other vegetable crops was from 30 to 50% higher with the membrane than on the check plots, because of the greater available moisture. The cost of the method depends mainly upon the price of asphalt or plastic sheeting. The machinery and tractor costs are relatively small. The economic justification depends upon the value of bringing sandy fields into agricultural production. For raising market vegetables near an urban center the benefits might well exceed the costs, since the expected life of the underground membrane is many years.

* The firm of Finkel & Finkel, Consulting Engineers, has prepared technical studies of this system in several countries.

REFERENCES

1. American Society of Agricultural Engineers, Design, layout, construction and maintenance of terrace systems, *ASAE Yearb.*, 530, 1978.
2. **Buchta, H. G., Broberg, D. E., and Liggett, F. E.,** Flat channel terraces, *Trans. ASAE*, 571, 1969.
3. **Dominy, P. F. and Worley, L. D.,** Design and construction techniques for parallel terraces, *Trans. ASAE*, 580, 1966.
4. **Hauser, V. L. and Cox, M. B.,** Evaluation of Zingg conservation bench terraces, *Trans. ASAE*, 462, 1962.
5. **Hillel, D.,** Runoff Inducement in Arid Lands, Volcani Institute of Agricultural Research, Rehovoth, Israel, 1967.
6. **Thomas, D. B., Barber, R. G., and Moore, T. R.,** Terracing of cropland in low rainfall areas of Kenya, *J. Agric. Eng. Res.*, 25, 57, 1980.
7. **Wittmuss, H. D.,** Topographic modification of land for moisture entrapment, *Trans. ASAE*, 384, 1968.

Chapter 9

GULLEY CONTROL

Herman J. Finkel

I. INTRODUCTION

The gulley is the most obvious and spectacular form of soil erosion as it causes large, unsightly gashes in the landscape. Deep gullies which spread back to the watershed divide can destroy an entire area and are the ultimate form of soil erosion known as "badlands". Not all gullies are so ominous, but they are insidious in their growth potential, and hence must be brought under control in the earliest possible stages. They appear in several forms, each of which has different characteristics and requires different methods of control. The two principal types are the so-called V and U gullies.

II. V GULLIES

A. Description

This type of gulley is characterized by a V-shaped cross section. It generally appears on sloping fields. The longitudinal gradient of the channel is greater than the slope of the land, as shown in Figure 1. The erosion is in the form of downward cutting in the center of the channel, causing it to become deeper as well as to grow backward, i.e., up the slope. Because this type of gulley forms on the more hilly lands, the distance back to the watershed divide is generally shorter and the catchment area feeding the gulley is small. The lateral spacing between one gulley and the next is likewise small. Consequently, the discharge passing through the channel is not large, but the velocity of flow may be quite high. On undulating topography every draw or depression may be the beginning of a V-type gulley. These dissect the field and make contour cultivation difficult, if not impossible. If they are not controlled, the farmer will be forced to plow the land up and down the slope, parallel to the gullies, and thereby make the erosion problem even more severe. It is of utmost importance, therefore, to apply control measures as soon as the gulley formation becomes evident.

B. Control Measures

V-shaped gullies often develop from rill erosion, when the water is concentrated from several rills into one channel. The best way to avoid the formation of gullies is to protect the area by contour cultivation, strip cropping, and where necessary, terracing. However, once the gulley is formed, the following engineering and agronomic measures are required to control it.

1. Elimination

If the gulley is not deep in its early stages, it is possible to simply fill it in with soil scraped from the field, or brought from a borrow pit. The danger of the gulley reopening again should then be prevented by the use of a system of broad-based terraces which cross the filled-in gulley. Once the cut has reached somewhat larger proportions, however, this solution may be uneconomical because of the need for large quantities of soil for fill.

2. Diversion

If the gulley is not long, it may be possible to run a diversion channel around the head

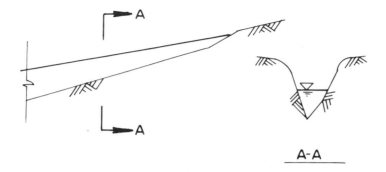

FIGURE 1.   Profile and cross section of V-type gulley.

of it to prevent surface runoff from entering while other control measures become established. The problem usually is, where to divert the water? It may be outletted to a more protected area of pasture, shrubs, or trees where it can flow away with less damage, while the cultivated field is being repaired. The water may also be diverted to an adjacent gulley which will indeed suffer more until the first gulley is healed. After this, the gulley which served as a temporary outlet for the diversion may then be protected by a diversion back to the first gulley. In other words, a gullied field is treated in stages with alternate gullies serving as outlets while those in between are being treated. If the gullies are close together, it may be possible to run a single diversion channel across the heads of all of them and discharge the water to some other area of lesser importance. This will permit the treatment of all the gullies simultaneously.

3. Check Dams

Even after the storm runoff is diverted away from the head of a gulley, there will still be some water collected from the sides. The control measures must take this into consideration. However, once the inflow has been reduced to a minimum, the next step is to prevent further downcutting, by reducing the velocity of flow in the center of the channel. This is best done by a series of small check dams built across the V-notch. The check dam may be any structure which retards the flow without actually stopping it completely. It is undesirable to build up a tight weir with an overfall, as this will cause a depression to be formed just below the weir.

The so-called "semi-permeable dam" may be made of plant residues such as branches, large leaves, straw, or stalks packed between a double row of wire mesh, and anchored firmly to the bed and sides of the gulley by either poles or stakes. Semi-permeable retardance structures may also be built of stone piled loosely in ridges across the channel, providing that the individual rocks are heavy enough to remain stable under flow. The height of these structures should be somewhat greater than the depth of flow as indicated by high water marks along the sides, but sufficiently below the top of the banks to pass the expected flow. The spacing of these checks along the length of the gulley has no precise formula and must be judged by experience. A practical range would be from 30 to 60 m (100 to 200 ft), varying inversely with the slope. Construction should begin at the upper end and proceed downstream. Although the retardance structures are permeable, there will, in the course of time, be some accumulation of soil behind each. The sharp apex of the V section will become broader and more rounded, reducing the hydraulic radius and the velocity of flow. When this happens, the progressive growth process will not only have been stopped, but to a certain extent reversed.

4. Treatment of the Sides

Once the downcutting has been stopped, the sides of the gulley may be stabilized by

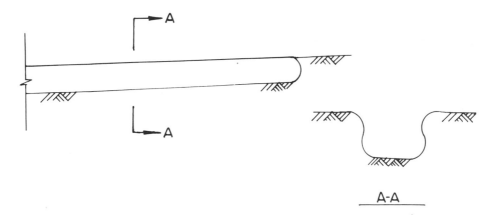

FIGURE 2. Profile and cross section of U-type gulley.

seeding or planting of perennial grass or vines. The species selected should be nonaggressive to the adjacent cropland, quick to become established, tolerant of low fertility in the subsoil, and have a strong root system. In certain regions a leguminous vine such as kudzu is used because of its high nutritive value as a fodder. The hay is cut by hand at regular intervals and fed to the animals.

III. U GULLIES

A. Description

The U-type gulley is recognized by its U-shaped cross section. The longitudinal slope of the channel bottom is usually parallel to the slope of the land through which it passes (Figure 2). It occurs on land with low slopes, almost approaching zero, and is often a source of surprise to travelers who do not expect to see such serious erosion on broad plains. In such flat topography the distance back to the watershed divide is relatively long, and the catchment areas are consequently large. The flood discharge passing through the U-type is larger, and the velocity of flow is usually slower than in the V-type. Runoff water enters the gulley from the head (upper end) and from the sides at points where the adjacent land may be slightly lower. In both cases the water flows over the vertical wall in a cascade and drops onto the flat channel bottom, forming a deep pool. This action causes undercutting and collapse of the bank. At the upper end, this collapse moves the head of the gulley upstream. The collapsing head may form branches, which in plan resemble cauliflower, as they broaden out. The collapsing sidewalls along the channel cause the entire section to become wider, but not deeper. Side gullies may branch out of the main channel at low points in the terrain. These, in turn, continue to grow headward. Though the spacing between gullies of this type is generally large, nevertheless the deep, long cuts through the field are a serious barrier to cultivation and require frequent bridges if left untreated. When the gulley begins to branch out in a hydra or multiheaded fan, and side gullies develop, the land is on its way to becoming ruined. Control measures should be undertaken before this stage is reached.

B. Control Measures

The active erosion of the U-type gulley is in the sidewalls and the headwall as a result of undercutting at the base of the vertical cut. The channel does not grow deeper, but becomes wider and grows in length headward. These are logical points of control. The methods employed are quite different from those described above for the V type. They should have the following sequence.

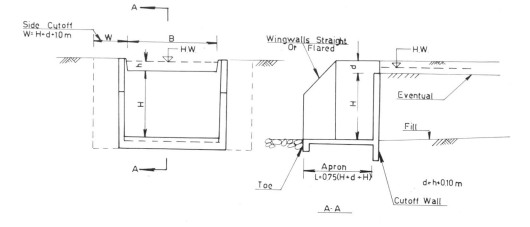

FIGURE 3. Permanent drop structure.

1. Raising the Datum

Since the longitudinal slope of the channel is small and parallel to the slope of the land, the most effective first step is to try to raise the baseline by means of a series of permanent, impermeable drop structures. The crest of the lowest weir should have an elevation only slightly lower than the apron of the structure above it. It should also be sufficiently below the height of the surrounding land to allow the design flow to pass the overflow section. In this way, transported soil will quickly accumulate behind the headwall almost to ground level, and taper off to the toe or apron of the structure above it. Thus, the lower reach of the channel between a pair of structures is completely filled in. The essential hydraulic parts of a drop structure are the weir, wing walls, apron, and toe cutoff wall, as shown in Figure 3. Construction details may be found in the extensive literature on soil conservation structures published by the U.S. Soil Conservation Service.

The structures may be made of concrete, stone masonry, steel sheet piling, or logs, according to availability and cost of these materials. In semiarid regions logs would generally be scarce, and in Third World countries both reinforced concrete and sheet piling would be expensive. Consequently, masonry built of field stone would probably be the material of choice.

An important factor in the design is the height and spacing of these drop structures. This may range from many low drops spaced close together, to fewer high drops spaced further apart. The limiting height of any drop structure is, of course, the depth of the gulley minus the hydraulic clearance required above the weir. Optimization studies have shown that within this limit, it is generally better to use fewer and higher drop structures.

Example:

Given: The channel of the gulley is 1.80-m deep and 2.5-m wide. The longitudinal slope is 0.005. The design flow is estimated, by the methods outlined in Chapter 2, to be 0.5 m³/sec.

Required: The dimensions and spacing of the drop structures.

Solution: Let us first calculate the overflow section required to handle the design flow. The formula for a broad-crested weir is approximately

$$Q = 1.7 \, B \, h^{1.5} \tag{44}$$

where B is the width and h is the head (both in meters) and Q is the discharge in m³/sec. Solving for h we get

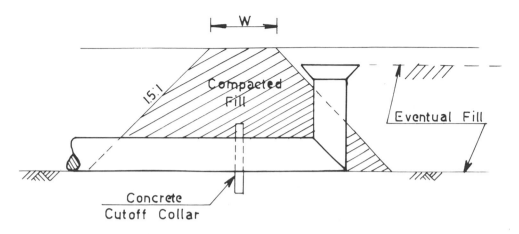

FIGURE 4. Tube and riser drop structure.

$$h = \left[\frac{Q}{1.7\ B}\right]^{2/3} = \left[\frac{0.5}{1.7 \times 2.25}\right]^{2/3} = 0.26 \text{ m} \qquad (45)$$

Note that the full channel width of 2.5 was reduced to 2.25 m weir width to allow for some thickness of the walls of the structure. If the depth of flow over the weir is 0.26 m, a depth of 0.30 should be used to allow a small amount of freeboard. The height of the drop through the structure, H, will therefore be $1.80 - 0.30 = 1.50$ m. The spacing of the drops along the channel will be $1.50/0.005 = 300$ m. This will create a level backfill of sediment material from the crest of the weir to the apron of the next higher structure. It is seen that the very low gradient of the channel in this type of gulley permits the structures to give effective control with a fairly wide spacing. This helps to offset their relatively high cost.*

Another type of drop structure is the earth fill with tube and riser shown in Figure 4. It is used where concrete is not available or too expensive. The tube and riser are laid in the channel and then covered with well-tamped earth. The slopes of the fill should be not steeper than 1.5:1 for the downstream side, and 1:1 for the upstream side. The latter can tolerate the steeper slope since it is expected to be filled with sediment after a short while. The top width of the fill need not be more than 1.50 m, but may be made wider to serve as a bridge for vehicles. The tube and riser may be made of poured concrete, precast concrete pipe, or steel pipe. In regions where these materials are too expensive, the tube and riser have been made from empty oil or tar drums welded together. A cutoff collar should be attached to the tube midway under the fill.

The hydraulic design of the riser is simply that of a sharp-crested weir, with the width, B, equal to the perimeter of the pipe. The formula is approximately

$$Q = 1.78 \text{ B } h^{1.5} \qquad (47)$$

The head, h, is the maximum water level over the crest of the riser. This crest must be at least H (height) below the surrounding ground elevation, plus some freeboard to avoid overtopping. It has been found that the H must equal or exceed the diameter of the riser to

---

* In the FPS system the solution to this problem is as follows:

$$Q = 17.5 \text{ cfs}; \quad B = 7.4 \text{ ft}$$

$$h = \left(\frac{17.5}{3.0 \times 7.4}\right)^{2/3} = 0.8 \text{ ft} \qquad (46)$$

have the pipe flow full. Since the limiting factor is the amount of water which can enter the riser, it is advisable to increase the length of the weir perimeter by using a funnel at the top of the riser. This permits more water to enter, and requires a smaller diameter tube and riser to carry this water away. At the lower end of the tube some form of energy dissipating structure should be built to prevent cutting at the toe of the fill. After construction, the lower face of the fill should be planted with grass to protect its surface from erosion.

2. Reshaping the Walls

After maximum control has been achieved by raising the datum with drop structures, the next step is to reshape the vertical sidewalls to a sloping bank not steeper than 1:1, but preferably flatter. This can be done with a bulldozer or a tractor-mounted power shovel. It can also be done by hand labor. The object is to prevent the sidewalls from being undermined, either by the flow in the bottom of the channel or by water entering from the sides as an overfall. The main work of reshaping will be in the reach of channel from just below a drop structure to halfway to the next lower structure. The reach above the drop structure should be largely filled in by sediment after the first few storms. It is interesting to note that the correct reshaping of a gulley channel is to give the U-type gulley a V cross section, and the V gulley a U cross section. Special care must be taken in the reshaping of the multiheaded overfalls at the upper end of the gulley to prevent further growth upstream.

3. Stabilizing the Channel

The reshaped section must now be stabilized by vegetative planting. Whereas in the smaller, more closely spaced V gullies, various types of grasses are used, the larger U gullies are generally stabilized by either vines or woody shrubs. There are varieties of shrubs and trees which form an excellent cover for the banks of large gullies. Some of these, especially the leguminous such as the carob or black locust, have edible pods and seeds which are valuable as fodder. Kudzu vines have also been used successfully for this purpose.

REFERENCE

1. **Little, W. C., Piest, R. F., and Robinson, A. R.,** SAE research program for channel stability and gulley control, *Trans. ASAE,* 362, 1980.

Chapter 10

WIND EROSION

Herman J. Finkel

I. INTRODUCTION

Wind erosion occurs mainly in the semiarid and subhumid climates and is one of the most serious hazards which threaten the agriculture of those regions. It is largely a manmade phenomenon resulting from the population explosion, which led to the destruction of the natural vegetative cover by clearing, burning, overgrazing, and plowing. It is one of the major causes of *desertification,* which is proceeding at an alarming rate in all continents except Europe.

In the U.S., the most critical region is the Great Plains. Kimberlin[11] reported that in the 1975/1976 season over 8 million acres (3.2 million ha) were damaged by wind erosion. Damage was reported for lands which lost 15 ton/acre (33.6 t/ha) or more. This is taken as the rate at which the erosion losses are visible to the eye. Damage, however, includes not only the physical loss of the topsoil, but reduction in the fertility and the water-holding capacity of the remaining soil. Similarly alarming statistics are reported for other parts of the U.S. and for the more agriculturally developed semiarid regions worldwide.

In addition to the permanent damage resulting from soil loss, wind erosion causes an annual economic damage from the destruction of crops. Many studies have been published which evaluate this type of loss. For example, Downes et al.[5] found a strong correlation between damage to young vegetable crops and a factor called the total kinetic effect (TKE), which is a single expression combining wind velocity, soil flux density, and time of exposure of the seedlings. Lyles[12] found that reduction in yields varied with the wind erodibility group (WEG) and ranged from 0 to 136 kg/ha/year for wheat and 204 kg/ha/year for sorghum (for a description of WEG, see Table 1). Reports of yield reductions due to wind storms abound in the technical literature of the highly developed agricultural regions. Less thoroughly reported are the more disastrous effects which occur in many of the underdeveloped regions, where wind storms can destroy a large portion of the subsistence crop of the small farmer.

In order to evaluate the potential hazard of wind erosion, and to select control measures which lie within the capacity of the farmers to establish, it is necessary to examine the physical nature of the erosion process.

II. THE WIND EROSION PROCESS

A. The Wind Causing Detachment and Transportation

The process whereby the wind detaches and transports soil particles has been the subject of intensive research during the past several decades, and a formidable number of scientific papers have been published on the subject. Many of these take as a starting point the fundamental work done by Bagnold[2] on the physics and mathematics of sand transport by the wind. An excellent review of the state-of-the-art may be found in Chepil and Woodruff.[4] The main principles of the wind erosion process will be briefly summarized here, as an understanding of them is essential for the selection and design of control measures.

When the wind blows across a rough ground surface the friction, or drag, reduces the velocity at the interface, and creates turbulence which exists up to a certain height above the ground. When the velocity of the wind is measured at different heights it is found to increase in proportion to the logarithm of the height, within the zone of turbulence. If the

Table 1
DESCRIPTION OF WEG[10]

WEG	Predominant soil, textural class	Dry soil aggregates >0.84 mm (%)	Soil erodability "I", mton/ha/year
1	Very fine, fine, and medium sands; dune sands	1	696
2	Loamy sands; loamy fine sands	10	301
3	Very fine sandy loams; fine sandy loams; sandy loams	25	193
4	Clays; silty clays; noncalcareous clay loams and silty clay loams with more than 35% clay content	25	193
4L	Calcareous loams and silt loams; calcareous clay loams and silty clay loams with less than 35% clay content	25	193
5	Noncalcareous loams and silty loams with less than 20% clay content; sandy clay loams; sandy clay	40	126
6	Noncalcareous loams and silty loams with more than 20% clay content; noncalcareous clay loams with less than 35% clay content	45	108
7	Silts; noncalcareous silty clay loams with less than 35% clay content	50	85

velocities are measured at two different heights and plotted on a semi-log graph as in Figure 1, a straight line connecting them will give the velocity at any other height. The slope of this line represents the velocity distribution of a specific wind, and is known as the drag velocity. If this line is prolonged, it will be found to intercept the Y-axis (or point of zero velocity) at a finite height, k, above the ground surface. For any other wind, having a straight line with a different slope of velocity distribution, the Y-intercept, k, will be the same. This value, k, is a function of the surface roughness, and represents approximately 1/30 of the height of the projecting particles.

The drag force or shear exerted by the wind on the surface of the ground (or vice versa) is given by the expression

$$\tau = \rho \, V_*^2 \tag{48}$$

where τ = the shear force or drag in g/cm^2, ρ = the density of air, taken at 1.22×10 g/cm^3, V_* = the drag velocity, which is also the tangent of the slope of the velocity distribution line.

From this relationship, and further experimental work it was found that the velocity of the wind at any height, z, is given by the following expression:

$$v = 5.75 \, V_* \log (z/k). \tag{49}$$

The terms k and V_* not only give a complete definition of the state of the wind, but also define the roughness of the ground surface. Thus, for example, if k = 0.015 cm, the roughness is $30 \times 0.015 = 0.45$ cm (see Figure 1).

This model of interaction between the wind velocity distribution, expressed by the drag velocity, and the rough ground surface, can be extended to the tops of the vegetative cover which form an irregular surface, restricting the movement of the wind. A surface of zero velocity is formed, known as the mean aerodynamic surface, which is usually somewhat lower than the maximum height of the vegetation as the tops of the plants permit some wind to pass through. The height of this surface above the ground is called Z_0 (Figure 2). The

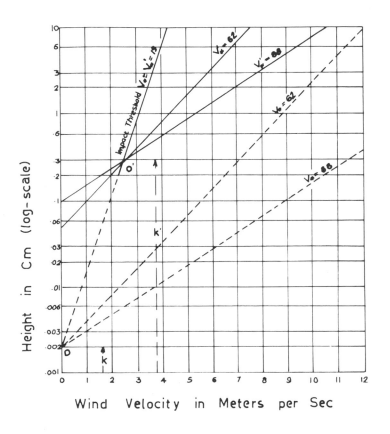

FIGURE 1. Distribution of wind velocity with height.[2]

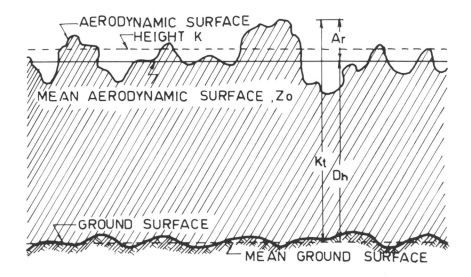

FIGURE 2. Wind over top surface of vegetation.[4]

value k expresses the average height above Z of the irregular surface connecting the tops of the plants. It is independent of the actual average height of the vegetation. Because of the much greater roughness of the vegetative cover, the expression in Equation 49 is modified to

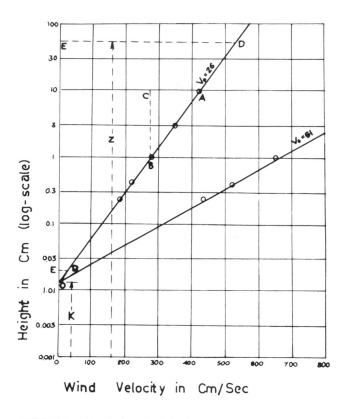

Wind   Velocity   in   Cm/Sec

FIGURE 3.   Distribution of wind velocity over an eroding surface.[2]

$$\tau_{avg} = a \rho V_*^2 \tag{50}$$

where a is a drag coefficient influenced by the height and kind of plant cover.

When the wind blows over a dry sandy field or other erodible soil surface it is found that the wind speed near the ground is reduced and the drag velocity has a different gradient than over a firm, noneroding surface. This is because the saltation of the detached particles creates a zone of interference within a certain height above the ground. All the different drag velocities on an eroding surface, each called V_*', pass through a common point $Z_0 +$ k' as shown in Figure 3. The velocity at this point, V_t, is a constant regardless of the strength of the wind and is actually the threshold velocity, or the velocity causing incipient detachment of the soil particle.

We can now write the expression

$$V_*' = \frac{v(z) - V_t}{5.75 \log \dfrac{z}{k'}} \tag{51}$$

This means that for a stronger wind, the velocity below the height k' is actually lower because more eroded soil particles are mixed into the air. This is also a function of the soil erodability. The greater the erodability, the greater the reduction of the wind velocity near the ground.

B. The Detachment of a Soil Particle

There is a certain similarity between the eroding of soil particles by the wind, and by flowing water which was described in Chapter 3. The same three principal causes are encountered. The first is the uplift force which detaches the particle and raises it vertically from the bed. This is caused by the so-called Bernoulli effect in which there is less pressure on the top of a grain as compared to the bottom, when the fluid velocity passing over the grain is increased. This is similar to the rise of water in the tube of an atomizer when a jet of air is blown across the top. Small, light grains of soil are detached in this way from the soil surface and they rise vertically into the airstream.

The rising particle is immediately subjected to a horizontal wind force which has a positive component on the windward side and a negative component on the leeward. Together they are known as the drag, which propels the grain forward. The combined path is a vertical rise and a forward thrust, which under the influence of gravity has a sloping, straight line descent meeting the soil surface at an angle of about 9°.

When the particle meets the surface, one of two things may happen. (1) It may strike heavier or well-anchored particles, rebound vertically into the airstream, and move forward again by saltation. In this case it will jump much higher into the wind than it did previously by uplift. It is, in fact, the myriad saltating particles which fill the wind for a substantial height above the ground and increase the previously referred to velocity drag. (2) The descending particle may, on the other hand, strike a loose particle on the ground and transfer its momentum. This other particle becomes detached by the process called abrasion, and enters the windstream as eroding material.

Some of the loose particles on the surface may receive an impact from the saltating particles which have a largely horizontal component. This will cause the grains to move or roll forward on the ground as surface creep. A saltating grain can, in this manner, move a grain which has six times its diameter. Consequently, the saltating material is generally finer, and the surface creep particles are coarser.

C. The Transportation of Soil by the Wind

There is an important distinction between the capacity of a given wind to transport eroded material, and the flux or quantity of material that wind will actually pick up and carry under given field conditions. The first is the upper limit of capacity of the wind to move a given type and size of grain, assuming an unlimited supply, and no inhibiting conditions. It depends upon the wind, as defined by the velocity gradient, and the eroding material as defined by average diameter, specific gravity, and degree of uniformity. Bagnold gives the following expression for dune sand:

$$q = C \, (d/D)^{0.5} \, \rho/g \, V_*^3 \tag{52}$$

where q is the rate of sand flow in grams per centimeter of lane width per second, d is the mean grain diameter, D is a standard grain diameter of 0.025 cm, ρ is the density of the quartz (2.65), and the velocity gradient or drag velocity is as previously defined. The constant, C, varies from 1.5 for a very uniform sand to 2.8 for a sand which has a very wide range of grain size.

Lyles[12] gives a somewhat different expression for particle flux as a function of the grain diameter and the drag velocity, each to some empirical exponent. As an alternative to the drag velocity, he suggests using the mean wind speed at some reference height, to an empirical negative exponent. (The constants and exponents will vary considerably with the soil type and field conditions.) This latter form was also proposed by Bagnold for the flow of average dune sand:

$$q = 1.5 \times 10^{-9} \, (v - V_t)^3 \quad \text{(in C.G.S. units)} \tag{53}$$

where v is the mean velocity at a height of 1 m and the other terms are as previously defined.

This simply means that the particle flux varies as the velocity at a standard height less the threshold velocity, raised to the third power. Most of the particle flux is in the form of saltation and surface creep, with 75% of the material moved by saltation. Unfortunately, for various agricultural soil types and more complex field conditions, the relationships are not as simple as those for dry, uniform dune sand.

There is a third form of transport by wind which merits attention. It is suspension. Dust carried by turbulent eddy currents may be carried to great heights and over long distances, as in the high, dense clouds which burgeoned over the U.S. Great Plains to create the disastrous Dust Bowl of the 1930s. Dust is carried by the wind when the drag velocity V_* is at least seven times the fall velocity of the particles. The quantity of dust carried will be directly proportional to the cube of the drag velocity and inverse to the diameter of the particle. It is interesting, however, that particles less than 0.02 mm in diameter are highly resistant to movement by the direct force of the wind. If fine dust lies on a smooth surface it will not be moved even by a very strong wind, because the roughness of the projections is minimal and the drag velocity consequently low. Bagnold has noted paths in the desert completely covered with a layer of dust which remained immobile in a high wind until it was stirred up by the passing cattle. In a soil of mixed texture the fine dust will cling to larger particles and thus enter the airstream, after which they separate. Once in the wind the dust has a great effect on the drag velocity. In this respect, particle suspension in air is very different from that in water. The ratio of the particle density to the fluid density is 1.65 in water and about 2000 in air. In other words, the transfer of momentum between the particle and the fluid involves a much greater volume of air than of water.

III. SOIL FACTORS AFFECTING RATE OF EROSION

A. General

As has been stated above, the capacity of a given wind to detach and transport eroded material is not the same as the amount actually eroded by the same wind under given conditions. The erosive capacity of the wind is modified, among other things, by the erodability of the soil. This, in turn, is influenced by soil texture, structure, moisture, organic matter, and lime. These will now be considered.

B. Soil Texture and Structure

These two factors must be considered together since the soil in a field (other than dune sand) is rarely in a loose, dry, unaggregated state. Chepil and Woodruff[4] state that the erodability index, I_w, varies with the amount of clay according to the following equation:

$$I_w = a\ G^b\ c^G \tag{54}$$

where G is the percentage by weight of clay (smaller than 0.002 mm) in the soil, and a, b, and c are constants. Most of the nonerodible clods are formed with the silt fraction (0.002 to 0.005 mm). Consequently, the higher percentage of this fraction, the lower the erodability. A high percentage of sand also increases erodability. The hardest clods (least erodible) are those formed with soils having 20 to 30% clay, 40 to 50% silt, and 20 to 40% sand.

Subsequent studies by others have attempted to quantify the erodability of different soil textural classes. Hayes[10] gives the WEG of soils under two cropping systems with stubble mulching in Texas (Table 1). It can be seen that the actual rate of erosion in metric tons per hectare per year varies inversely with the percent of dry soil aggregates. Of course, the soil texture, like the wind, is a given condition which is difficult, if not impossible, to change. Nevertheless, it is valuable to be able to compare the potential wind erosion hazard on different soils, for purposes of land-use planning and resource allocation for control measures.

C. Soil Moisture

Water has a very important influence on the erodability of a soil in several ways. Adsorbed moisture forms in a film around individual soil particles. The cohesion between these films creates a resistance to detachment which the force of the wind must overcome.

Wetting and drying cycles cause the soil to form water-stable aggregates which are held together by insoluble cementing materials. These aggregates are very slow to disintegrate. Having a greater equivalent diameter than the individual particles, they have a greater drag force. To be resistant to wind erosion and yet have good workability for agriculture, a soil should have a substantial proportion of water-stable aggregates larger than 1 mm.

Falling raindrops tend to disperse and sort the soil material, leaving the fines on the surface. Upon drying, a crust is formed which is resistant to wind erosion. This crust may be only a few millimeters thick. The mechanical stability of the crust increases with the percentage of particles smaller than 0.02 mm which are dispersable in water. The medium-textured soils such as silt loams and silty clay loams have the highest tendency to form crust and are hence more resistant to wind erosion. Both sandy loams and clays are less likely to form protective crusts under raindrop splash.

On the other hand, excessive soil moisture followed by the weathering processes of wetting and drying, freezing and thawing, and high winds can have the effect of breaking down the soil aggregates which are not water stable. The disintegrated material leaves the fine particles exposed to wind erosion in the dry season. Examples of this process are presented by Gillette[9] as analyzed by scanning electron micrographs.

It can be seen that the presence of soil moisture in almost any form is an aid to reducing wind erosion. Previous action by falling rain or flowing water, followed by a drought period with high winds, will increase wind erosion for certain soil textures. These conclusions emphasize the seriousness of the wind erosion problem in the semiarid regions. If the fields could all be irrigated the hazard would be greatly reduced. Barring this as economically unfeasible in many regions, maximum attention should be paid to water conservation and water harvesting, not only for the sake of growing crops, but also to help conserve the soil.

D. Organic Matter

There is a tendency to assume that soil conditions which inhibit water erosion will also be helpful against wind erosion, but in the case of organic matter this is not entirely true. Studies carried out in Canada and the U.S. show that when organic matter is in the process of rapid decomposition in the soil there is an increase in cloddiness and water-stable aggregation. This is caused by various cementing substances produced by the microorganisms which attack the vegetable matter. These substances are decomposition products of plant residues, secretory products of the microorganisms, such as slime, gums, and mucus, and polysaccharides synthesized by the microorganisms. However, the soil aggregation so produced is temporary, and begins to be broken up by the action of other forces into friable particles which are erodible by the wind. The time required for the disintegration of the aggregates varies with the original amount of the vegetative matter and its rate of decomposition. The remaining humus is highly erodible in the dry season. Consequently, the long-term effect of a one-time plowing under of organic matter is negative, insofar as erosion is concerned. It is recommended, therefore, that crop residues should be left on the surface to decompose more slowly so that the cementing substances will be replenished for a longer period of time. If the vegetative matter is anchored to the surface it will provide additional protection from the wind.

E. Calcium Carbonate

Chepil and Woodruff[4] report that additions of lime in the form of precipitated calcium carbonate decreased the amount and stability of clods in the soil, and hence increased erodability. This was true for all soil textures tested except on a highly erodible loamy sand, where the opposite effect was noted. In soils with a high content of humus, the lime had a very pronounced effect of decreasing water-stable aggregates. In semiarid regions there is often a layer of accumulated calcium carbonate just below the plow-depth. When this is turned up and left on the surface it has a deleterious effect on wind erosion. Moreover, since the erosion is more pronounced on hilltops, the calcium carbonate layer is first exposed there and carried by the wind to the lower parts of the field, where it generally increases the erodability of the soil.

IV. CONTROL MEASURES

A. Introduction

Wind causes three general types of damage: loss of soil through erosion, harm to crops, and contamination of the atmosphere by dust. There are many measures which can be taken to reduce the severity of these types of damages. They may be divided into two broad categories: reduction of the force of the wind, and protection of the surface of the soil. Some measures, however, produce both effects simultaneously, and hence the two categories must be considered together.

B. Vegetative Cover

As has been seen in the theoretical treatment of wind erosion, a vegetative cover raises the aerodynamic surface of zero velocity to some height above the ground and reduces the drag on the surface particles. Vegetative cover can provide varying degrees of wind erosion control up to a maximum of almost 100% depending upon the nature, density, and orientation of the crop. Perennial grasses and legumes provide almost complete protection, once they are established; however, if they are subsequently overgrazed, the degree of protection will rapidly diminish. In semiarid regions susceptible to wind erosion it is important to select a rotation which provides the greatest possible number of years of perennial pasture consistent with the economic needs of the farmers. This means, of course, combining animal husbandry with agriculture, a step which is not always acceptable to the rural population (for a treatment of this problem, see Chapter 4). Once established, the pastures should be carefully managed with respect to their carrying capacity so as to avoid overgrazing.

Close-growing grain crops are next in the hierarchy of effectiveness, but only after the stand is reasonably high. In the agricultural calendar of a semiarid region it will be found that there are only a few months between the time the grain is high enough to afford wind protection, and the time of harvest. This is the only period when the crop is effective. If this period is followed by a dry season with high winds, the growing grain crop will have provided very little protection.

Row crops, such as maize, sorghum, and cotton are less effective than grain because of the wide, bare spaces between the rows; however, not all row crops are equal in this respect. Maize tends to have a cloddy soil surface between the rows and is better protection than some of the well-tilled vegetable crops. The date of planting is also important. Some crops such as sorghum and cotton in the midwestern U.S. come to maturity after the season of severe winds has passed. It is important, therefore, to compare the agricultural calendar with the known seasons of wind erosion to estimate the degree of protection which the selected crops may offer the soil. However, the opposite problem of wind damage to the crops is no less important. If a crop can be grown in the season of less wind, it should be done so to realize maximum yields. The soil protection, in this case, would have to be accomplished by other measures.

C. The Field

The size, shape, and orientation of the field all have an influence on the erodibility of the soil and wind damage to the crops.

Table 2
RECOMMENDED
AVERAGE WIDTHS FOR
WIND STRIP CROPPING
IN THE GREAT PLAINS
REGION

	Width	
Soil texture	ft	m
Sand	20	6
Loamy sand	25	7.5
Granulated clay	80	24
Sandy loam	100	30
Silty clay	150	45
Loam	250	75
Silt loam	280	85
Clay loam	350	100
Silty clay loam	430	130

Considering the orientation, there is considerable importance to the angle which the crop rows make with respect to the direction of the prevailing wind. It should be as close to a right angle as possible. Siddoway, in unpublished data quoted by Chepil and Woodruff,[4] measured the erosion in a wheat field with the rows 25-cm apart. The erosion was six times greater when the rows were parallel to the wind direction than when at right angles. Woodruff and Zingg[15] found that with sorghum stubble in rows 1-m apart in wind tunnel tests, the soil erosion was three times greater with the rows parallel to the wind as compared to perpendicular. The lesson is obvious, but can only be applied in the field when the erosive wind is known to have a consistent prevailing direction. A certain amount of deviation can, of course, be tolerated but this should not exceed 45° to either side. If the wind rose shows greater variation, additional control measures should be considered.

The dimension of the field in the direction of the prevailing wind should be limited. In small holdings or a fragmented land tenure pattern, this is rarely a problem unless, by coincidence, many adjacent farmers plant their fields to the same crop with the same orientation. However, in the large grain farms of the U.S. Great Plains, the Ukranian Wheat Belt, or the Pampas of Argentina, the fields may stretch as far as the eye can see. The best way to limit the width of the field is to plant the area in strips, normal to the prevailing wind. On these strips a rotation can be grown such as wheat — sorghum — fallow, or better still, meadow — meadow — wheat — fallow. The meadow can be a grass or biennial legume. This implies a combined grain and animal husbandry enterprise, with the attendant difficulties discussed previously. The widths of the strips, as recommended by Woodruff et al.,[14] are given in Table 2.

A problem, typical of semiarid regions, may arise in fields with a rolling topography subject to water erosion during the rainy season, in addition to the wind erosion of the dry season. There is a high degree of probability that the contour strips would not necessarily be oriented correctly with respect to the wind direction. There is no simple solution to this dilemma, and a compromise must be sought in terms of greater protection for the greater hazard, assuming that the relative hazards of the two types of erosion can be evaluated. However, in large, extensive fields a partial solution is possible. The strip cropping should be laid out on the contour, and at certain lengths along the strip the crops should be alternated. This gives a sort of checkerboard pattern which shortens the dimension of any one crop parallel to the wind, no matter from which direction it blows.

Table 3
AVERAGE EFFECTS OF KIND AND ORIENTATION OF
CROP RESIDUE ON EROSION AND SANDY LOAM SOIL
BY WIND OF UNIFORM VELOCITY[a]

| | Quantity of soil eroded in a wind tunnel | | | |
| | Covered with wheat residue | | Covered with sorghum residue | |
Quantity of crop residue above soil surface (lb/acre)	Standing, 10-in. high (ton/acre)	Flat (ton/acre)	Standing, 10-in. high (ton/acre)	Flat (ton/acre)
0	16.0	16.0	16.0	16.0
500	2.8	8.5	13.0	14.5
1,000	0.1	2.5	8.1	10.4
2,000	T[b]	0.1	3.9	5.3
3,000	T	T	1.4	2.2
6,000	T	T	T	0.2

[a]   Unpublished data from F. H. Siddoway.
[b]   T = trace, insignificant.

D. Crop Residues and Minimum Tillage

Once the crop has been harvested the field will be exposed to accelerated wind erosion unless additional protection is provided. One of the most effective measures is to leave the crop residues on the field. Stubble mulch refers to the residues left standing, whereas straw mulch means that the residue after corn picking or grain combining is spread flat on the field. Wind tunnel studies by Siddoway, reported by Chepil and Woodruff,[4] are given in Table 3. The data show a decided advantage to standing residue over flat, and wheat residue over sorghum. With 2 ton/acre (4.5 t/ha) of crop residue, almost all of the wind erosion was controlled. Stubble mulch farming requires that throughout the year the operations of tillage, planting, cultivating, and harvesting be handled in such a way as to leave maximum coverage on the ground of either growing crops or plant residues. This requires the use of special tillage equipment, subsoiling for under-surface cultivation, and planting equipment designed to operate in a trash-covered field.

In addition to other farming practices which maintain surface residues, minimum tillage should be practiced. The various systems of minimum tillage were discussed in Chapter 4. Experience with minimum tillage systems for wind erosion control in the Great Plains of the U.S. is summarized by Fenester and Wicks.[16] These systems are used where wheat is grown every other year, alternating with fallow, to accumulate moisture. There is a conflict between clean cultivation to control weeds, which would otherwise steal the moisture, and the wind erosion, which would be severe on a clean cultivated fallow. The method giving the best results was subtillage with large V sweeps. This killed the weeds without removing the plant residues on the surface. The annual soil loss from wind erosion was 2.0 t/ha as compared to 6.5 t/ha from the plots which were plowed and cultivated with conventional equipment. The methods and benefits of stubble mulch fallow and minimum tillage apply equally to water and wind erosion.

E. Mulches

Mulching generally refers to the covering of the soil surface with material brought in from outside the field. It may be used for control of both water and wind erosion, moisture conservation, weed control, temperature modification, and many other purposes. The subject of mulches for wind erosion control is well summarized in Armbrust[1] from whom some of the following material is adapted.

There is a wide variety of materials which may serve as mulch, including crop residues, gravel, asphalt, plastic sheets, and various synthetic organic and inorganic products. The principal properties of a good mulch are the following:

1. Indispersible in water
2. Durable enough to last for at least several seasons
3. Porous enough to allow percolation of rainfall and sprouting of seedlings
4. Application without difficulty
5. Relatively inexpensive

Vegetative mulches not actually grown in a field may be brought in and spread, but they must be given adequate anchorage against blowing away. Chopped hay or straw, for example, can be disked into the ground and subsequently packed down with some type of corrugated roller in which the disks are spaced from 10 to 20 cm apart and penetrate the soil about 6 cm. It may also be anchored by mixing with liquid asphalt, either cutback or emulsified. The straw is chopped and spread with a blower, to which nozzles are attached for adding the correct amount of asphalt. When the thinning solvent evaporates, or the emulsion breaks, a semi-solid layer of asphalt and grass is established on the surface. These methods avoid the need for heating the asphalt in the field. Rates of application adequate for control of erosion under a 38.0 m/sec wind were 5 to 6 t/ha of straw or hay with 3 to 6.6 m^3/ha of asphalt.

Workers have had some success in the direct application of asphalt alone by spraying through long-range nozzles. This method has been used extensively in Iran to control erosion and stabilize sand dunes prior to planting of trees and grasses. It has not yet been applied there to other cropland. Of course, the economics of using a pure asphalt mulch over vast tracts of land could be favorable only in a petroleum producing country.

A wide variety of synthetic substances for mulching has been tried in the U.S. These include liquid plastics, resins, asphalt emulsions, latex compounds, and many others. Comparative effectiveness and costs may be found in Armbrust.[1] At this time, however, the broad conclusions are that these products are for the most part too expensive. If this is the conclusion for the U.S. how much more so will it apply to Third World countries which would have to import these products. The economic advantage still lies entirely with vegetative residues produced on the site.

F. Barriers

Anchoring the soil surface to reduce its erodability is one side of the equation. Reducing the velocity, and hence the erosiveness of the wind, is the other side. This is done principally with barriers or, as they are called, windbreaks. These may be made of different materials ranging from a solid, impervious wall, to an open line of trees or shrubs. The height, spacing, orientation, width, and porosity of windbreaks have all been the subject of extensive research in the wind tunnel (theoretical) and in the open field (practical). From these studies a number of generalizations have been arrived at, including the items in the following four paragraphs.

The effectiveness of barriers is expressed in terms of the percentage reduction in either wind velocity or erosiveness at distances of multiples of H to the leeward of the barrier, where H is the height of the barrier. The relative erosiveness is the better criterion for the present purpose. It is taken as the velocity cubed, as long as the range of velocities observed is well above the V_t or threshold velocity. If, for example, a certain type of barrier reduces the velocity by 25% at, for example, 10 H the relative velocity will be 75%. The relative erosiveness of the wind at this point will be 42%, which means a reduction in relative erosiveness of 68% from the value to the windward of the barrier. In general, 20 to 30 times the height should be taken as the maximum spacing unless experimental data is available locally to revise this recommendation.

A solid barrier provides the maximum reduction in wind erosiveness just in its own lee, but the effect extends for only a short distance. Several investigators found that the greatest overall wind speed reduction between a barrier and 30 H was with a barrier having 40% porosity. This is now accepted as good design practice.

The greatest effect of wind erosiveness reduction is achieved with a barrier oriented at 90° to the wind direction. This effectiveness is rapidly reduced as the angle diverges from normal. In areas where the wind direction is variable, it is necessary to crisscross the fields with barriers in both directions.

The most effective cross-sectional shape of the windbreak has a triangular windward surface, rather than an abrupt vertical barrier. This can be achieved by planting several rows of different trees and shrubs, with the low-growing species on the exposed side. The former tendency to plant a wide shelterbelt of as many as ten rows of trees is now being revised to narrower stands of two or three rows. The greater porosity gives effective protection with much less loss of land, and with lower cost. An even more important factor to consider in the semiarid regions is the competition for soil moisture between the windbreak and the crop which is being protected. Because of root spread, the nonirrigated, rain fed crop will suffer near the windbreak. For many common species of trees, the root spread is estimated to be 2.5 H. It is greater for deciduous and less for conifers. Consequently, narrow windbreaks of conifers would be the windbreak of choice. For various regions in the U.S. the Soil Conservation Service has published lists of the recommended species and the distance between the barriers. In the developing countries of the Third World such lists must be worked out from experience in each region. The trend should be toward locally available species which can be established during the rainy season with no supplementary irrigation, and which will be hardy enough to withstand drought. Narrow shelterbelts should be favored to save water, land, and expense.

REFERENCES

1. **Armbrust, D. V.,** A review of mulches to control wind erosion, *Trans. ASAE,* 904, 1977.
2. **Bagnold, B. A.,** *The Physics of Blown Sand and Desert Dunes,* Methuen & Co., London, 1941.
3. **Biswas, M. R.,** U.N. conference on desertifiction in retrospect, *Environ. Conserv.,* 5(4), 1978.
4. **Chepil, W. S. and Woodruff, N. P.,** The physics of wind erosion and its control, *Adv. Agron.,* 15, 211, 1965.
5. **Downes, J. D., Fryrear, D. W., Wilson, R. L., and Sabota, C. M.,** Influence of wind erosion on growing plants, *Trans. ASAE,* 885, 1977.
6. **Fryrear, D. W. and Downes, J. D.,** Consider the plant in planning wind erosion systems, *Trans. ASAE,* 1070, 1975.
7. **Fryrear, D. W. and Wiegand, C. L.,** Evaluating wind erosion from aerial photos, *Trans. ASAE,* 892, 1977.
8. **Fryrear, D. W. and Lyles, L.,** Wind erosion research accomplishments and needs, *Trans. ASAE,* 916, 1977.
9. **Gillette, D. A.,** Fine particulate emissions due to wind erosion, *Trans. ASAE,* 880, 1977.
10. **Hayes, W. A.,** Designing Wind Erosion Control Systems in the Midwest Region, Tech. Note LI-9, Soil Conservation Service, U.S. Department of Agriculture, Lincoln, Neb., 1972.
11. **Kimberlin, L. W., Hidlebaugh, A. L., and Grunewald, A. R.,** The potential wind erosion problem in the United States, *Trans. ASAE,* 873, 1977.
12. **Lyles, L.,** Wind erosion: processes and effect on soil productivity, *Trans. ASAE,* 880, 1977.
13. **Skidmore, E. L. and Hagen, L. J.,** Reducing wind erosion with barriers, *Trans. ASAE,* 911, 1977.
14. **Woodruff, N. P., Lyles, L., Siddoway, F. H., and Fryrear, D. W.,** How to Control Wind Erosion, Agric. Info. Bull. No. 354, U.S. Department of Agriculture, Washington, D.C., 1972.
15. **Woodruff, N. P. and Zingg, A. W.,** Soil Conservation Service Tech. Pap. No. 112, U.S. Department of Agriculture, Washington, D.C., 1952.
16. **Fenester, C. R. and Wicks, G. A.,** Minimum tillage fallow systems for reducing wind erosion, *Trans. ASAE,* 906, 1977.

INDEX

Z